NATIONAL GEOGRAPHIC
SCIENCE

FLORIDA
SCIENCE INQUIRY AND WRITING BOOK

School Publishing

PROGRAM AUTHORS

Judith S. Lederman, Ph.D.
Randy Bell, Ph.D.
Malcolm B. Butler, Ph.D.
Kathy Cabe Trundle, Ph.D.
David W. Moore, Ph.D.

Science Inquiry

Life Science

Life Science

CHAPTER 3 How Do Plants and Animals Respond to Seasons?

Earth Science

CHAPTER 4 What Properties Can You Observe About the Sun?

Science Inquiry

Earth Science

CHAPTER 5 **What Can You Observe About Stars?**

Physical Science

How Can You Describe and Measure Matter?

Physical Science

Science Inquiry

Physical Science

CHAPTER 9

What Is Light?

Science in a Snap!

Next Generation Sunshine State Standards **SC.3.L.14.2** Investigate and describe how plants respond to stimuli (heat, light, gravity), such as the way plant stems grow toward light and their roots grow downward in response to gravity. (Also **SC.3.N.1.6**)

Science in a Snap! Observe a Plant's Growth

Place a plant in a box that has a hole cut in one side. Close the box and put it in the sun. After several days, **observe** the plant. Rotate the plant and close the box. **Predict** how the plant will grow. After several days, observe the plant again. How has the plant grown? **Infer** why it is growing in that direction.

Next Generation Sunshine State Standards **SC.3.N.1.1** Raise questions about the natural world, investigate them individually and in teams through free exploration and systematic investigations, and generate appropriate explanations based on those explorations. (Also **SC.3.L.15.1**)

CHAPTER 2

Science in a Snap! Compare Animals

Observe the animals in the pictures. **Compare** their body parts. Find the following parts on one or more of the animals:

- Fur
- Feathers
- Scales

Classify these animals based on these coverings.

Next Generation Sunshine State Standards **SC.3.L.17.1** Describe how animals and plants respond to changing seasons.

CHAPTER 3

Science in a Snap! Count Hibernation Heartbeats

When an animal hibernates, its heart beats less often. For example, when a woodchuck is not hibernating, its heart beats about 80 times in a minute. When the woodchuck hibernates, its heart beats very slowly. Use a stopwatch. Make a fist and squeeze your fingers together tightly 1 time every 15 seconds. That's how often a woodchuck's heart beats when it is hibernating. How many times does the hibernating woodchuck's heart beat in 1 minute? How might a slower heartbeat help a woodchuck survive when it hibernates?

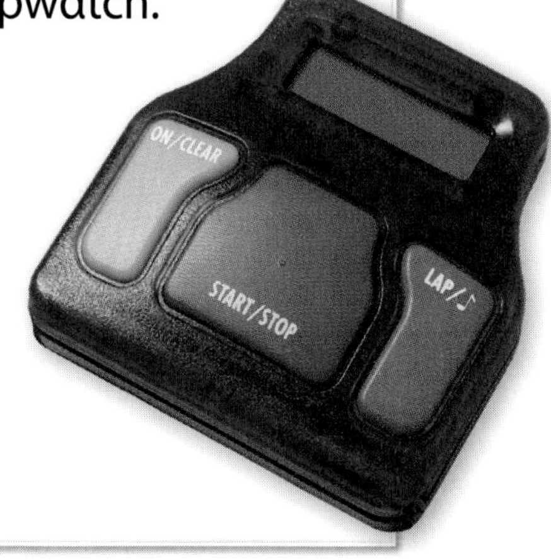

Explore Activity

Investigate Plants and Gravity

How does gravity affect the growth of plant roots?

Science Process Vocabulary

observe verb

When you **observe,** you use your senses to learn about an object or event.

predict verb

When you say what you think will happen, you **predict.**

Materials

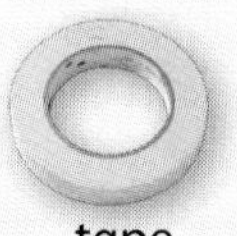

tape

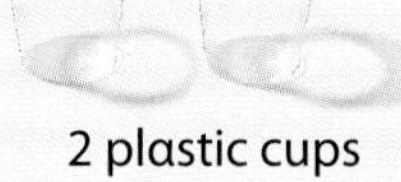

2 plastic cups

8 paper towels

2 bean seeds

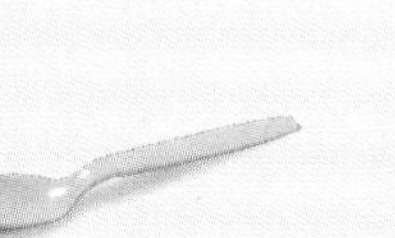

spoon

water

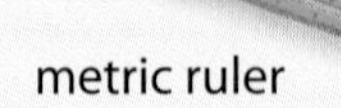

metric ruler

clay

Next Generation Sunshine State Standards SC.3.L.14.2 Investigate and describe how plants respond to stimuli (heat, light, gravity), such as the way plant stems grow toward light and their roots grow downward in response to gravity. (Also **SC.3.N.1.3, SC.3.N.1.6**)

What to Do

1 Use tape to label the cups **A** and **B.** Bunch up 4 paper towels in each cup. Place a bean seed in each cup right next to the side of the cup so you will be able to see its roots grow. Use a spoon to water the seeds. Add enough water to wet the whole paper towel.

2 Every other day, water the seeds. Add the same amount of water to each cup. **Observe** the seeds, watching for them to sprout. Record your observations in your science notebook.

What to Do, continued

3. When the roots start to grow, wait until they are about 1.5 cm in length. Then carefully lay cup A on its side. Use the clay to hold cup A in place. Do not change the position of cup B. **Predict** in which direction the roots of each plant will grow. Record your predictions.

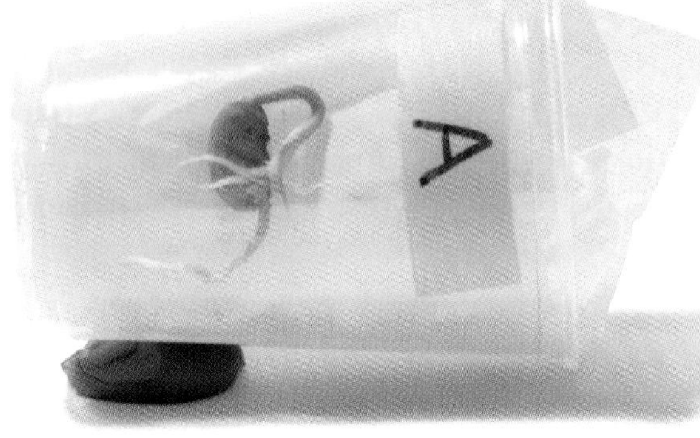

4. Continue to observe and water the plants every other day. When the roots have grown another 1.5 cm, carefully turn cup A upright. Do not change the position of cup B. Predict how the roots will grow. Record your prediction.

5. After 10 days, observe the roots in each cup. Record your observations.

Record

Write and draw in your science notebook.
Use a table like this one.

How Plant Roots Grow

Day	Cup A Observations	Cup B Observations

Explain and Conclude

1. Do the results support your **predictions?** Use your **observations** to explain.
2. **Infer** the way gravity affects the growth of plant roots. Tell what evidence from this activity you observed to make your inference.

Parts of the roots of these cypress trees grow above water in this swamp.

Directed Inquiry

Investigate Plant Parts

How can you classify plants by their parts?

Science Process Vocabulary

observe verb

You can **observe** objects or events by using one or more of your five senses.

classify verb

When you **classify,** you put things in groups according to their characteristics.

I can classify the plants by looking for things they have in common.

Materials

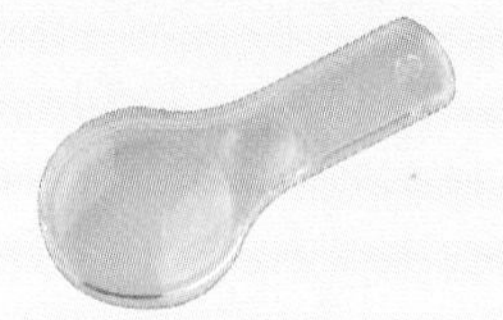

hand lens

sunflower seed

corn seed

rye grass seed

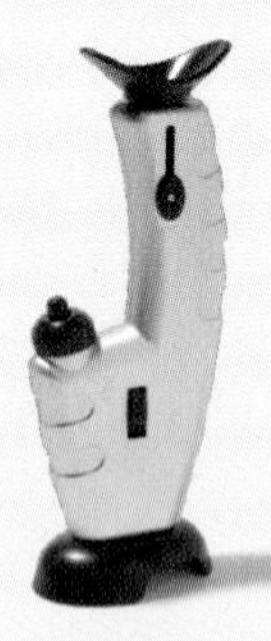

microscope

microscope slide with spores

What to Do

1. **Observe** and draw the plants in the photos. Look for plant parts such as roots, stems, leaves, and flowers. Record your observations in your science notebook.

Sunflower

Rye grass

Corn

What to Do, continued

2 The sunflower, rye grass, and corn plants produce seeds. Use a hand lens and microscope to observe the 3 different kinds of seeds. Record your observations and draw each seed.

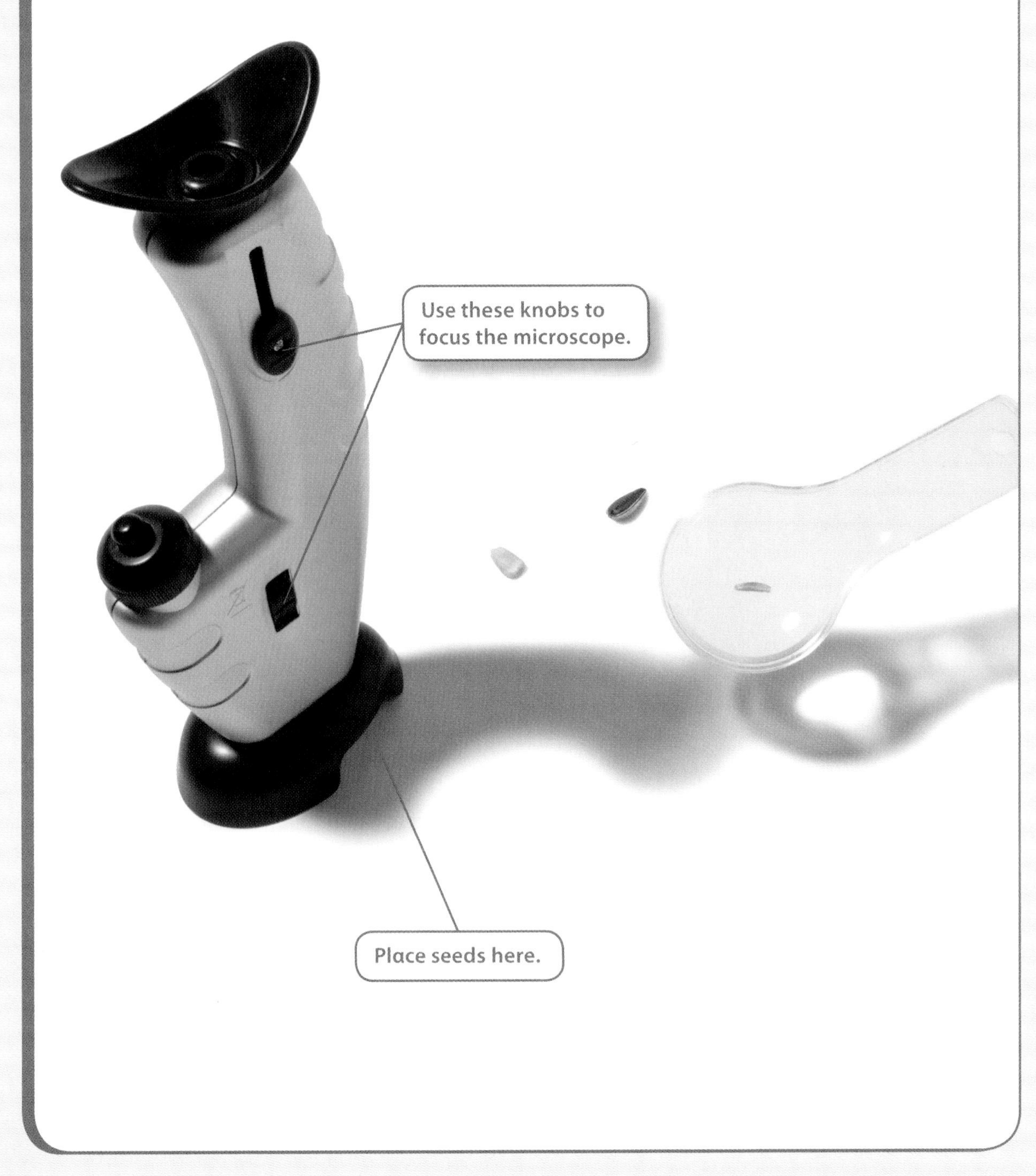

What to Do, continued

3 Observe the moss and the ferns in the photos. These plants do not produce seeds. They produce spores instead. Record your observations and draw the plants.

Boston fern

Moss

Bracken fern

What to Do, continued

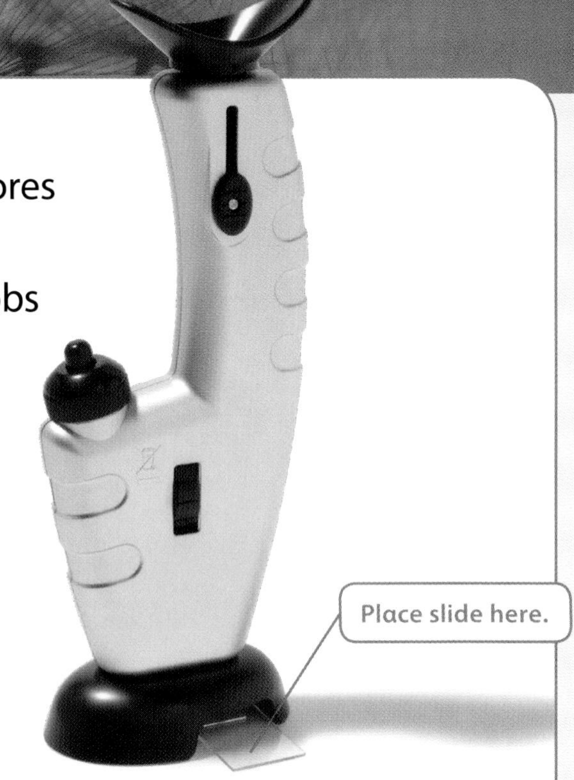

4. Place the microscope slide with spores under the microscope. Turn on the microscope's light, and use the knobs to focus it. Observe the spores on the slide. Draw your observations.

5. Look at the plant pictures on pages 15 and 17. Review your drawings in your science notebook. Review your observations of the seeds and the spores on the slide.

6. Use your observations of the seeds and spores to **classify** the plants in the photos below as either seed-producers or spore-producers. Share your classification with other groups. Think about other ways that you could classify the plants in the photos.

A

B

C

Record

Write and draw in your science notebook.
Use tables like these.

Observations of Plants and Plant Parts

	Draw Your Observations
Sunflower	
Rye grass	

Observations of Seeds and Spores

	Draw Your Observations
Corn seed	
Sunflower seed	
Rye grass seed	
Spore slide	

Explain and Conclude

1. How are the plants in the photographs on pages 15 and 17 similar and different?
2. How did your **observations** of seeds and spores help you to **classify** the plants in the photos on page 18?

Think of Another Question

What else would you like to find out about the parts of plants?

Next Generation Sunshine State Standards
SC.3.N.1.3 Keep records as appropriate, such as pictorial, written, or simple charts and graphs, of investigations conducted.

Math in Science

Using Tables to Organize Information

When you do science research or a science investigation, you collect information. A table can make understanding information easier. A table is especially helpful when collecting numerical data or when comparing objects.

The table shows the kinds of sea turtles that are found in the Florida Panhandle. Using the table, you can quickly find the scientific name and common name of each kind of sea turtle.

Sea Turtles of the Florida Panhandle

Scientific Name	Common Name
Caretta caretta	Atlantic loggerhead sea turtle
Chelonia mydas	Atlantic green sea turtle
Lepidochelys kempii	Kemp's ridley sea turtle

Atlantic loggerhead

Atlantic green

Kemp's ridley

Collecting and Organizing Information Erin was researching information about the different kinds of animals that live in the Florida Panhandle. She found the following information on a reliable Web site.

> The Florida Panhandle has more than 788 different kinds of animals with backbones. These include fish, amphibians, reptiles, birds, and mammals.

Erin decided to organize the information in a table.

Animals in the Florida Panhandle

Kind of Animal
Fish
Amphibians
Reptiles
Birds
Mammals

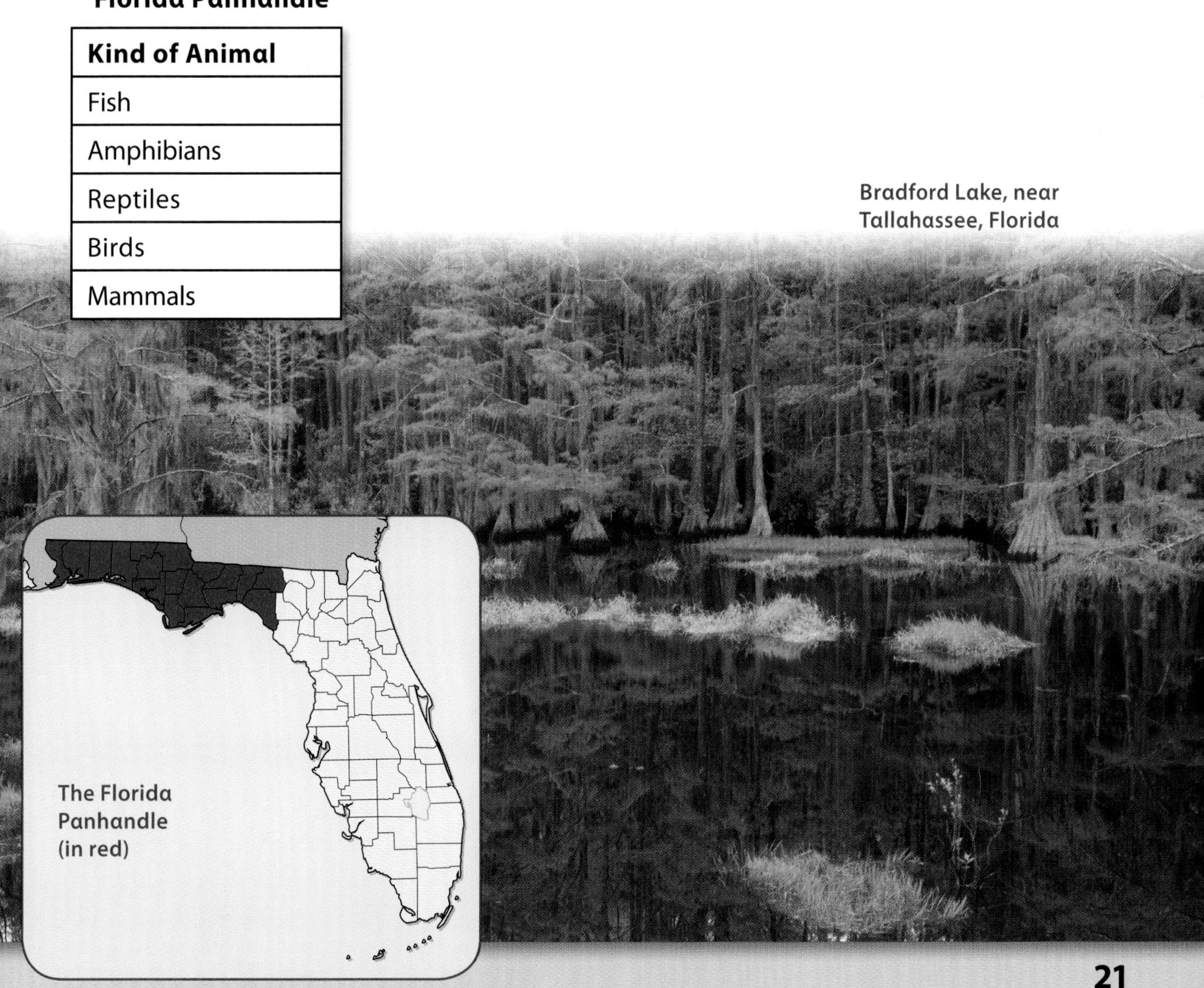

Bradford Lake, near Tallahassee, Florida

The Florida Panhandle (in red)

Next Erin wondered what the characteristics were of each kind of animal. She did some more research. She added a column to her table for the new information.

Animals in the Florida Panhandle

Kind of Animal	Characteristics
Fish	Live in water. Have fins and a tail. Breathe with gills.
Amphibians	Live part of their lives in water and part on land.
Reptiles	Lay eggs on land. Hard dry scales cover their body. Most live on land.
Birds	Have two wings, a beak, and feathers. Lay eggs.
Mammals	Have hair or fur. Give birth to live young. Females make milk for their young to drink.

SUMMARIZE

What Did You Find Out?

1. Why is using a table to record information helpful?

2. With what two kinds of information is using a table especially helpful?

Practice Organizing Information

Erin continued to look for information about the Florida Panhandle. Here is what she found.

> The Florida Panhandle has more different kinds of organisms than almost any other place in the United States. No area of its size in the United States has more kinds of frogs or snakes. In fact, the Panhandle has 52 kinds of amphibians and 80 different kinds of reptiles. It also has more than 200 kinds of fish. Scientists have identified 57 kinds of mammals and more than 400 kinds of birds.

Copy the table from page 22. Then add the new information Erin collected to the table.

Directed Inquiry

Investigate Animal Classification

How can you identify animals with backbones and model how a backbone works?

Science Process Vocabulary

classify verb

When you **classify,** you put things in groups according to their characteristics.

I can classify animals by observing how they are alike and different.

model noun

You can use a **model** to show how something in real life works.

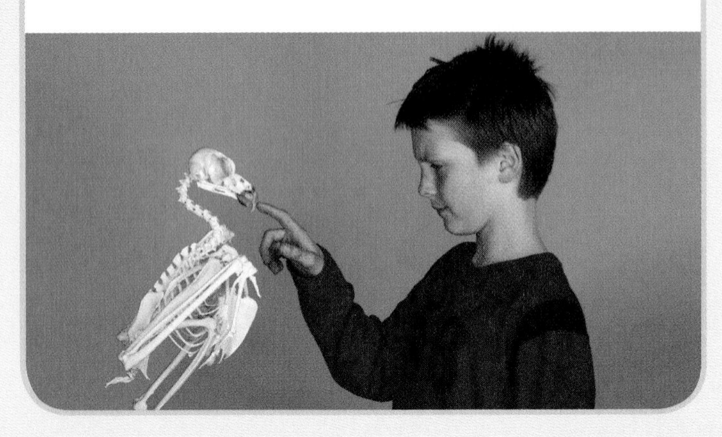

Materials

Animal Picture Cards

chenille stem

wooden beads

metal washers

Next Generation Sunshine State Standards **SC.3.L.15.1** Classify animals into major groups (mammals, birds, reptiles, amphibians, fish, arthropods; vertebrates and invertebrates; those having live births and those which lay eggs) according to their physical characteristics and behaviors. (Also **SC.3.N.1.1, SC.3.N.1.3, SC.3.N.1.6**)

What to Do

1. **Observe** the animals on the Animal Picture Cards. Look at each animal's body parts, body coverings, and other physical features. Record your observations in your science notebook.

2. **Classify** the animals into 2 groups: animals that have backbones and animals that do not have backbones. Record your classifications.

What to Do, continued

3 You will make a **model** of a backbone. Study the picture to see the parts of a dog's backbone.

The backbone is made of many small bones. The spaces and discs in between each bone allow the backbone to move.

4 Choose one of the animals with a backbone from the Picture Cards. Use the chenille stems, beads, and metal washers to make a model of its backbone.

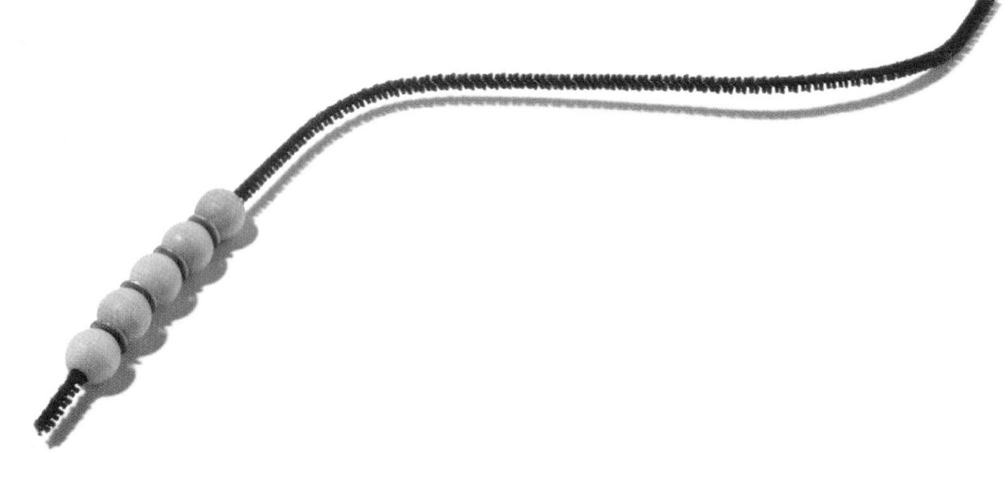

5 Observe the beads and washers as you bend and twist your model. Record your observations.

Record

Write and draw in your science notebook.
Use a table like this one.

Animal Classification

Animal	Observations	Backbone or No Backbone?
Chimpanzee		
Clownfish		

Explain and Conclude

1. How did you use your **observations** to **classify** the animals as animals with backbones and animals without backbones?
2. **Infer** how the structure of the backbone helps an animal move different ways. Use your observations of the **model** backbone to explain your answer.

Think of Another Question

What else would you like to find out about animals with backbones and how their backbones work? How could you find an answer to this new question?

Tyrannosaurus rex

Guided Inquiry

Investigate Arthropods

How can you classify some common arthropods?

Science Process Vocabulary

observe verb

When you **observe,** you use your senses to learn about an object or event.

I can see that this spider has 2 main body parts.

classify verb

When you **classify,** you put things in groups according to their characteristics.

Materials

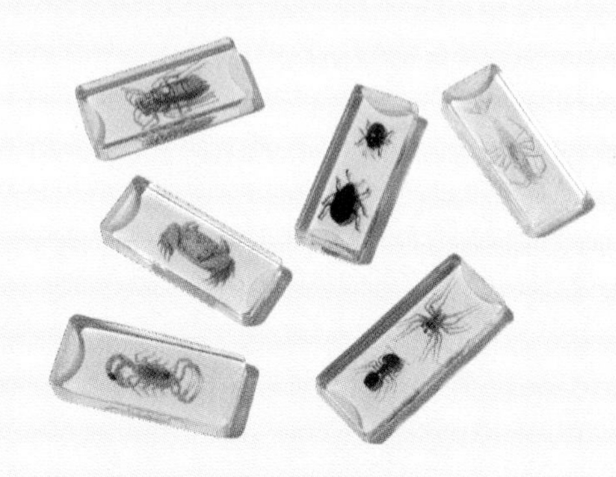

arthropod blocks

ruler

hand lens

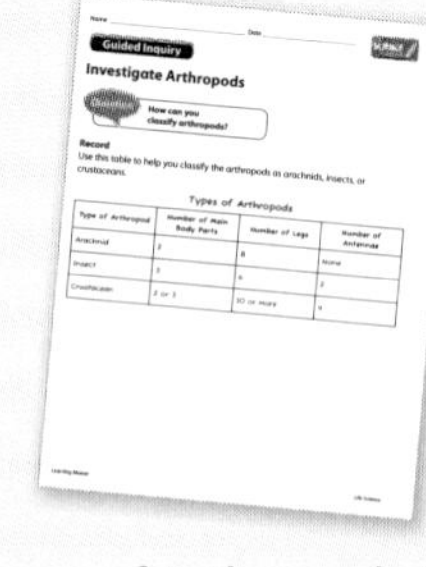

Name ______ Date ______

Guided Inquiry

Investigate Arthropods

How can you classify arthropods?

Record

Use this table to help you classify the arthropods as arachnids, insects, or crustaceans.

Types of Arthropods

Type of Arthropod	Number of Main Body Parts	Number of Legs	Number of Antennae
Arachnid	2	8	None
Insect	3	6	2
Crustacean	2 or 3	10 or more	4

Types of Arthropods table

Next Generation Sunshine State Standards **SC.3.L.15.1** Classify animals into major groups (mammals, birds, reptiles, amphibians, fish, arthropods; vertebrates and invertebrates; those having live births and those which lay eggs) according to their physical characteristics and behaviors. (Also **SC.3.N.1.1, SC.3.N.1.3**)

What to Do

1. Select an arthropod. Draw it in your science notebook. **Measure** its length. Record your measurements.

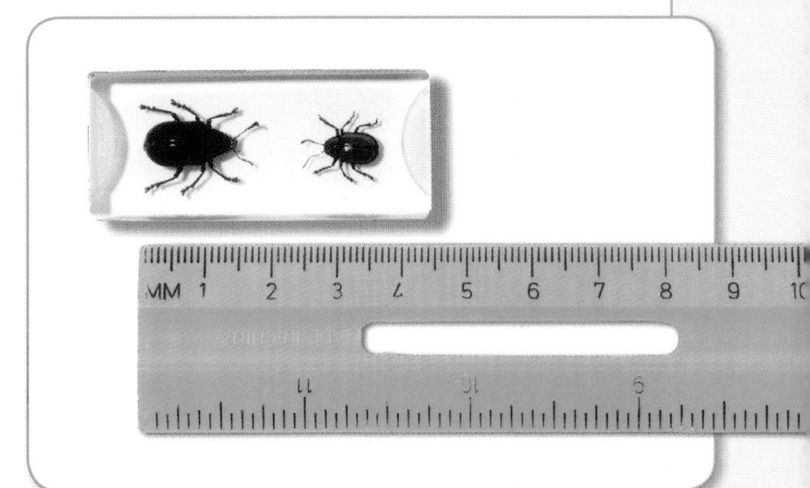

2. **Observe** the arthropod with the hand lens. **Count** the main body parts that your arthropod has. Record the number.

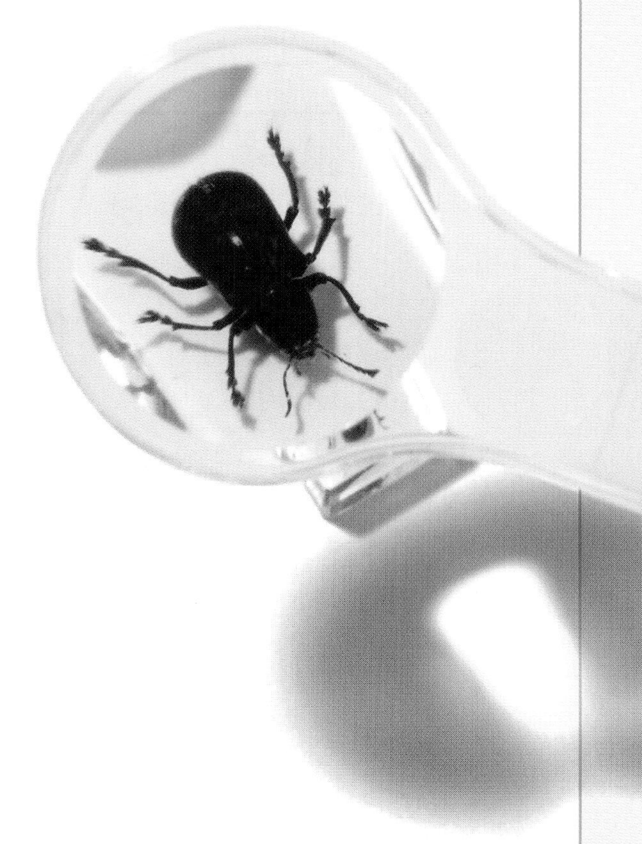

3. Observe the arthropod again. Count the number of legs it has. Turn the block over if necessary. Record the number.

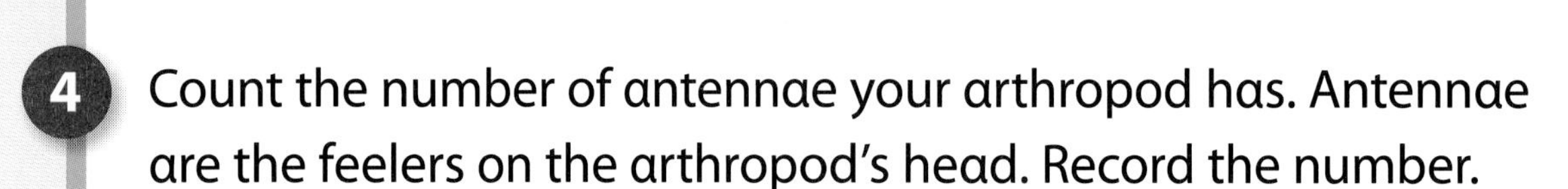

What to Do, continued

4 Count the number of antennae your arthropod has. Antennae are the feelers on the arthropod's head. Record the number.

5 Use the Types of Arthropods table to **classify** the arthropod as an arachnid, insect, or crustacean. Record your classification.

Name ________ Date ________

Guided Inquiry

my SCIENCE notebook

Investigate Arthropods

Question: How can you classify arthropods?

Record

Use this table to help you classify the arthropods as arachnids, insects, or crustaceans.

Types of Arthropods

Type of Arthropod	Number of Main Body Parts	Number of Legs	Number of Antennae
Arachnid	2	8	None
Insect	3	6	2
Crustacean	2 or 3	10 or more	4

Learning Master

Life Science

6 Observe 2 more arthropods by trading with other groups. Repeat steps 1–5 with each arthropod. **Compare** your classifications with other groups.

Record

Write and draw in your science notebook.
Use a table like this one.

Arthropod Information

Name of arthropod			
Drawing			
Length (cm)			
Number of main body parts			
Number of legs			

Explain and Conclude

1. What type of arthropods did you observe? How do you know?
2. How else could you **classify** the arthropods?

Think of Another Question

What else would you like to find out about classifying arthropods? How could you find an answer to this new question?

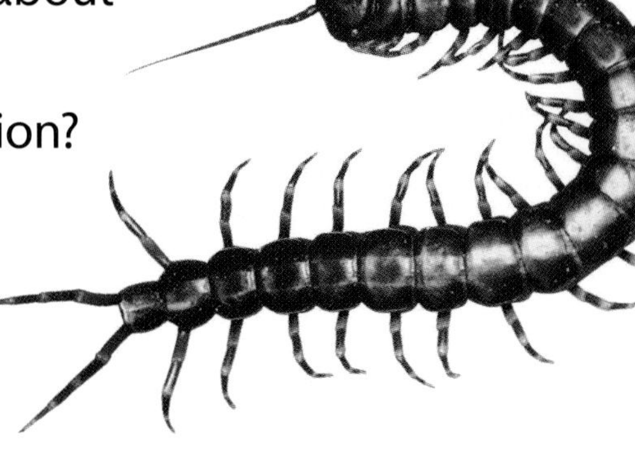

Millipedes and centipedes are arthropods, too.

Directed Inquiry

Investigate Temperature and Cricket Behavior

How does temperature affect cricket behavior?

Science Process Vocabulary

data noun

Data are observations and information that you collect and record in an investigation.

share verb

When you **share** results, you tell or show what you have learned.

Materials

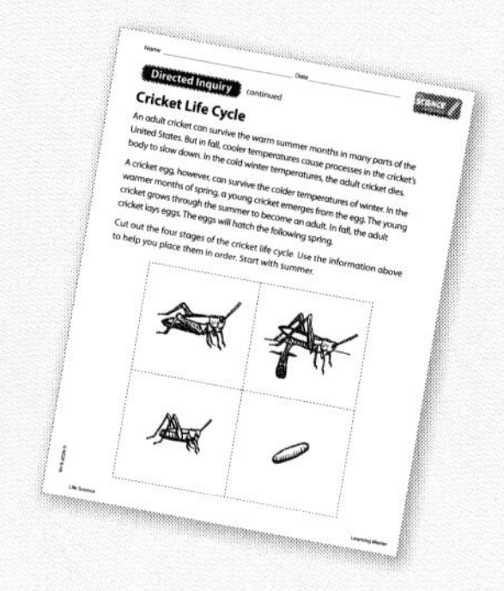

Name ____ Date ____

Directed Inquiry continued

Cricket Life Cycle

An adult cricket can survive the warm summer months in many parts of the United States. But in fall, cooler temperatures cause processes in the cricket's body to slow down. In the cold winter temperatures, the adult cricket dies.

A cricket egg, however, can survive the colder temperatures of winter. In the warmer months of spring, a young cricket emerges from the egg. The young cricket grows through the summer to become an adult. In fall, the adult cricket lays eggs. The eggs will hatch the following spring.

Cut out the four stages of the cricket life cycle. Use the information above to help you place them in order. Start with summer.

Stages in the Life of a Cricket chart

scissors

cricket habitat

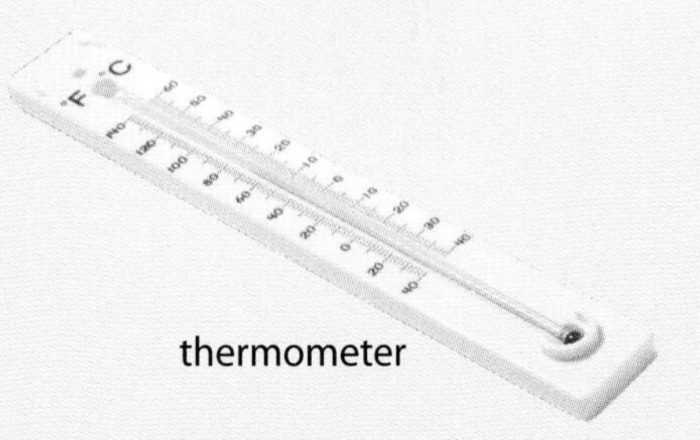

thermometer

Next Generation Sunshine State Standards **SC.3.L.17.1** Describe how animals and plants respond to changing seasons. (Also **SC.3.N.1.1, SC.3.N.1.3, SC.3.N.1.6**)

What to Do

1. Read the information on the Stages in the Life of a Cricket chart. Cut out the pictures and put them in order to show the cricket's life stages.

2. You can **observe** how temperature affects adult crickets by observing their behavior at different temperatures. **Measure** the temperature of the cricket habitat. Record the **data** in your science notebook.

3. Observe the activity of the crickets in their habitat. Pay special attention to how active they are. Record your observations.

What to Do, continued

4 Use the thermometer to measure the temperature in a refrigerator. Record your data.

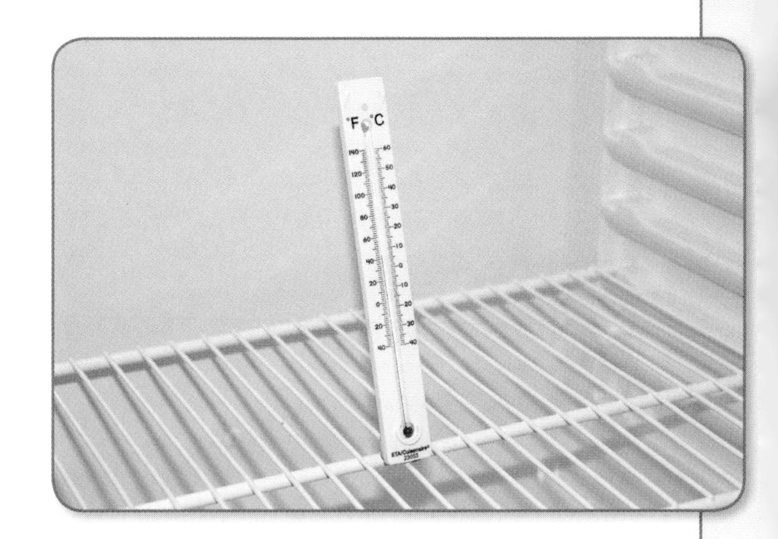

5 Place the cricket habitat in the refrigerator. **Predict** how the cooler temperature will affect the crickets' behavior. Record your predictions.

6 Remove the cricket habitat from the refrigerator after 10 minutes. Observe the cooled crickets in their habitat. Record your observations.

7 After 5 minutes record the temperature again. Observe and record the crickets' behavior.

Record

Write in your science notebook.
Use a table like this one.

Cricket Activity

Where Habitat Was	Temperature (°C)	Observations
Classroom		
Refrigerator (after 10 minutes)		

Explain and Conclude

1. **Share** your results with the class. Did your results support your **predictions?** Explain.
2. How did the change in temperature affect the crickets' behavior?
3. When the temperature of its environment gets cooler, the processes in a cricket's body slow down. **Infer** how this might affect a cricket's ability to move and survive.

Think of Another Question

What else would you like to find out about how temperature affects crickets' behavior? How could you find an answer to this new question?

This red cricket lives in wetland habitats.

Guided Inquiry

Investigate Temperature and Seed Sprouting

Science Process Vocabulary

variable noun

A **variable** is something that can change in an experiment.

In this experiment, I will change only one variable.

conclude verb

You **conclude** when you use information, or data, from an investigation to come up with a decision or answer.

Based on the results, I can conclude how temperature affects the growth of the seeds.

Materials

2 resealable plastic bags

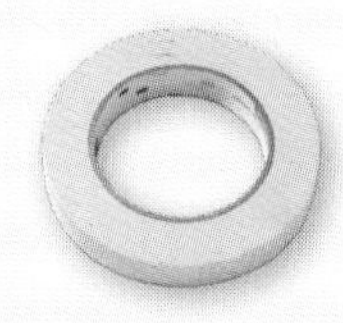
tape

2 paper towels

spray bottle with water

seeds

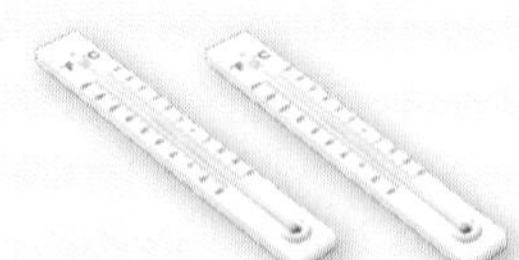
2 thermometers

Do an Experiment

Write your plan in your science notebook.

Make a Hypothesis

In this investigation, you will grow seeds at different temperatures.
How will temperature affect seed sprouting?
Write your **hypothesis.**

Identify, Manipulate, and Control Variables

Which variable will you change?
Which variable will you observe or measure?
Which variables will you keep the same?

What to Do

1. Label the plastic bags **1** and **2.** Fold the paper towels and place one in each bag. Use the spray bottle to make the paper towels damp. Spray both paper towels the same number of times. Then put a thermometer in each bag.

2. Choose which type of seed you will test. Place 1 seed on the paper towel in each bag. Seal the bags.

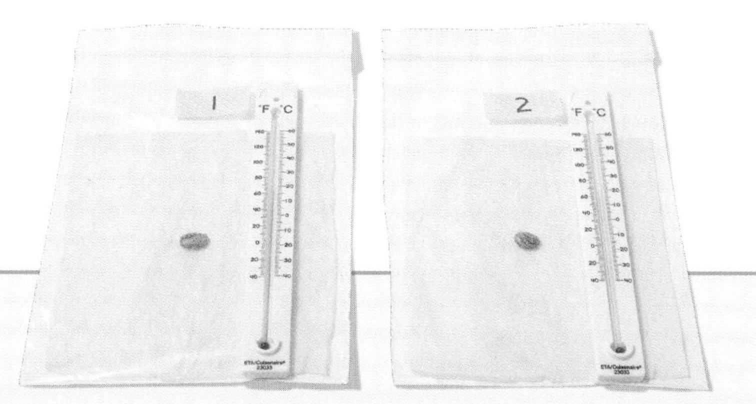

What to Do, continued

3. Place bag 1 on a flat surface in a dark place in the room. After 30 minutes, use the thermometer to **measure** the temperature. Record the **data** in your science notebook.

4. Decide whether you will put bag 2 in the refrigerator or the freezer. Place the bag on a flat surface in the place you choose. Wait 30 minutes and then measure the temperature of the bag. Record your data.

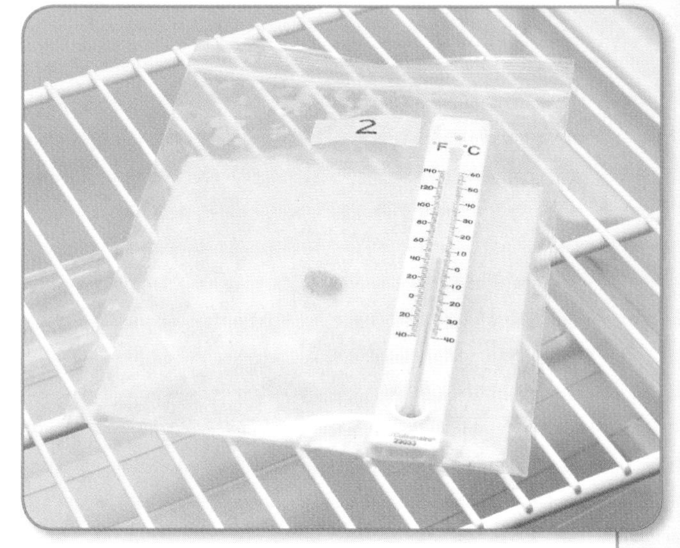

5. **Observe** both bags every other day for a week. Look for signs that your seeds are sprouting. Record your observations.

Record

Write and draw in your science notebook.
Use a table like this one.

Temperature and Seed Sprouting

Day	Observations	
	Bag 1: ____ °C	Bag 2: ____ °C
1		
2		

Explain and Conclude

1. Which seed had grown more by the end of the week?
2. What can you **conclude** about temperature and how seeds sprout?
3. How might seed growth be affected in places that have cold winters?

Think of Another Question

What else would you like to find out about temperature and seeds sprouting?

Open Inquiry

Do Your Own Investigation

Question **Choose one of these questions, or make up one of your own to do your investigation.**

- How do light and darkness affect how seeds sprout and grow?
- How does increasing the amount of light affect the growth of a plant?
- How does temperature affect how long it takes for a butterfly pupa to change to an adult?
- Are ants attracted to a light environment or a dark environment?

Science Process Vocabulary

hypothesis noun

You make a **hypothesis** when you state a possible answer to a question that can be tested by an experiment.

I will test the hypothesis that cold temperatures slow down the growth of an insect.

Open Inquiry Checklist

Here is a checklist you can use when you investigate.

Next Generation Sunshine State Standards SC.3.N.1.1 Raise questions about the natural world, investigate them individually and in teams through free exploration and systematic investigations, and generate appropriate explanations based on those explorations. (Also **SC.3.N.1.3**)

- ☐ Choose a **question** or make up one of your own.
- ☐ Gather the materials you will use.
- ☐ If needed, make a **hypothesis** or a **prediction.**
- ☐ If needed, identify, manipulate, and control **variables.**
- ☐ Make a **plan** for your **investigation.**
- ☐ Carry out your **plan.**
- ☐ Collect and record **data. Analyze** your data.
- ☐ Explain and **share** your results.
- ☐ Tell what you **conclude.**
- ☐ Think of another question.

Four Peaks Mountain, Sonora Desert, Arizona

Desert plants have adaptations that help them to survive in a very dry environment.

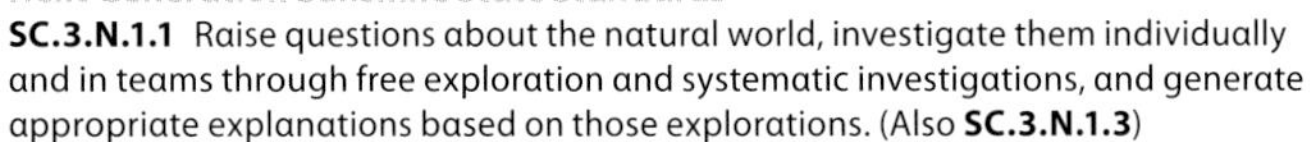

Next Generation Sunshine State Standards
SC.3.N.1.1 Raise questions about the natural world, investigate them individually and in teams through free exploration and systematic investigations, and generate appropriate explanations based on those explorations. (Also **SC.3.N.1.3**)

Write About an Investigation

Observing Insect Behavior

The following pages show how one student, Tess, wrote about an investigation. As she read about living things and their environments, Tess became interested in finding out whether insects prefer a light or dark environment. Here is what she thought about to get started:

Some ants build large ant hills like this one.

- Tess wanted to work with living things that she could observe in the classroom, so she decided to use ants in her investigation.
- She needed to use materials that she could obtain easily and work with safely. She would base her question on the materials she could use and the environment she would create.
- Tess decided to observe 2 ant habitats, one in light and one in darkness, to see to which one the ants are attracted.
- She would observe the ants to see whether more ants were in the light habitat or the dark habitat.

Model

Question

Will ants be attracted to a light environment or a dark environment?

Make sure the question asks exactly what you are trying to find out.

Materials

2 ant habitats, same kind, same size, with ants

connector tube

hand lens

dark cloth

Carefully describe the objects and materials you will need to answer your question. Be specific about the materials you will use.

Your Investigation

Now it's your turn to do your investigation and write about it. Write about the following checklist items in your science notebook.

- [ ] **Choose a question or make up one of your own.**
- [ ] **Gather the materials you will use.**

Model

My Hypothesis

If I place 1 ant habitat in the light and another in the dark, then most of the ants will move to the darker habitat. If I leave both habitats in the light, then there will be about the same number of ants in each habitat.

State what you think will happen in your experiment. You can use an "If..., then...." statement to make your hypothesis clear.

Your Investigation

☐ **If needed, make a hypothesis or prediction.**

Write your hypothesis or prediction in your science notebook.

Model

Variable I Will Change

One habitat will be in a light environment.
The other will be in a dark environment.

Variable I Will Observe

I will observe to see if more ants move
to the light or dark habitat.

During an investigation, it is important to accurately record observations.

Variables I Will Keep the Same

Everything else will be the same. Both habitats will be the same kind and size. They will contain the same type and amount of food and water. Both will be placed in the same location.

Your Investigation

☐ **If needed, identify, manipulate, and control variables.**

Write about the variables in your investigation.

Model

My Plan

1. Label the habitats **Light** and **Dark**.
   Connect the habitats with the connector tube.
2. Observe the ants with a hand lens.
   Record the physical characteristics of the ants.
3. Observe the 2 habitats. Estimate which habitat has more ants. Record your estimate.
4. Cover the Dark habitat with the dark cloth.
   Leave the habitats in a light place overnight.
5. The next day, observe the habitats. Look under the cloth to estimate the number of ants. Record the number of ants in each habitat.
6. Repeat step 5 each day for a week.
   Record your estimates.

Your plan should tell everything you will do. Give exact times or amounts when necessary.

Your Investigation

☐ **Make a plan for your investigation.**

Write the steps for your plan.

Model

I carried out all 6 steps of my plan.

You might make a note if you needed to adjust your plan in any way. Tess found she did not need to adjust her plan.

Your Investigation

☐ **Carry out your plan.**

Be sure to follow your plan carefully.

Model

Data (My Observations)

Observations of Ant Habitats

Day	Light Habitat		Dark Habitat	
	Observation	Estimate	Observation	Estimate
1	The ants have 3 body parts, 2 antennae, and 6 legs. They are active.	There are about 100 ants in the habitat.	The ants have 3 body parts, 2 antennae, and 6 legs. They are active.	There are about 100 ants in the habitat.
2				

Use a table to organize your observations.

My Analysis

At the start of the experiment, both habitats had about the same number of ants. After 1 week, the Dark habitat had more ants than the Light habitat.

Explain what you observed during the entire investigation.

Your Investigation

☐ **Collect and record data. Analyze your data.**

Collect and record your data, and then write your analysis.

Model

How I Shared My Results

I took photos of the habitats every day. At the end of the week, I used the photos to make a computer presentation. I shared the presentation with classmates.

Using pictures, diagrams, and graphs can help others understand your results.

My Conclusion

The ants in the Light habitat moved to the Dark habitat when I covered it with a dark cloth. The results of the investigation support my hypothesis.

Make a conclusion based on your data and observations. Check whether your results support your hypothesis.

Another Question

I wonder what would happen if I placed one ant habitat in a warm environment and the other in a cold environment. Would the ants move to the warm habitat or the cold habitat?

One investigation often gives you ideas for other investigations.

Your Investigation

- Explain and share your results.
- Tell what you conclude.
- Think of another question.

Next Generation Sunshine State Standards
SC.3.N.1.6 Infer based on observation.

How Scientists Work

Inferring

Scientists make observations and collect data. Then scientists use the observations and what they already know to make an inference.

You, too, make inferences. For example, one morning you might observe that the weather is cloudy and windy. You know that it often rains when the weather is cloudy and windy. So, you infer that it might rain sometime soon. Your inference helped explain your observation.

You can use observations of the clouds and wind to infer that there will be a storm.

Making an Inference Scientists found this fossil in Wyoming. They identified it as the fossil of an alligator.

Scientists know that living alligators need warm, moist weather all year. They also found fossils of other animals and plants that need warm, moist weather in the same area of Wyoming. Today, Wyoming has warm summers and cold winters. Sometimes the weather is very dry. Scientists used the fossil information to infer that Wyoming's weather was once warmer and wetter than it is now.

▲ Fossil alligator

▼ Alligator today

Making More Inferences

More than 100 years ago, scientists discovered a fossil that they thought was an example of Earth's earliest bird. The *Archaeopteryx* fossil showed a creature that had feathers, a wishbone, a beak, and wings like birds today. It also had fingers, claws, and teeth like many dinosaurs.

The *Archaeopteryx* fossil was found over 100 years ago.

Scientists used what they observed about the fossil and what they knew about modern birds to infer what *Archaeopteryx* looked like.

Scientists inferred that the animal could fly like a bird, but it also behaved like a dinosaur in some ways. Because the *Archaeopteryx* is extinct, they cannot observe its behavior. The scientists made inferences based on what they observed in the fossil and what they knew about birds and dinosaurs.

What Did You Find Out?

1. What do scientists use to make inferences?

2. What evidence did scientists use to infer that the weather in Wyoming at one time was wetter and warmer than it is now?

Inferring

The island of Spitsbergen is in the Arctic Ocean, near the country of Norway. The climate of Spitsbergen is very cold for most of the year. A large part of the island is covered with ice. Scientists have discovered fossils of ferns and other tropical plants that lived on Spitsbergen millions of years ago. The fossils are similar to plants that today live in warm, tropical climates.

Scientists found fossils of ferns and other plants on the island.

Infer what the climate of Spitsbergen was millions of years ago. Explain your inference.

The climate of Spitsbergen today is very cold.

Science in a Snap!

Next Generation Sunshine State Standards **SC.3.E.5.2** Identify the Sun as a star that emits energy; some of it in the form of light. (Also **SC.3.E.6.1**)

CHAPTER 4

Science in a Snap! Compare Sunlight and Shade

Stand in the sunlight for 3 minutes. Tell how you feel and what you see. Move to the shade. Tell how you feel and what you see. **Compare** how you feel in the sunlight to how you feel in the shade. What causes the differences you feel and see?

Next Generation Sunshine State Standards SC.3.E.5.1 Explain that stars can be different: some are smaller, some are larger, and some appear brighter than others; all except the Sun are so far away that they look like points of light.

Science in a Snap! Observe Differences in Stars

Look at the picture of the stars. All stars except the sun look like points of light because they are so far away. Try to find some bright stars. Now look for faint, or dim, stars. Choose a section of the picture and **count** the number of different colored stars you can see. Explain how a large star could appear dimmer than a smaller star. Explain how 2 stars that are alike can appear to have different brightnesses.

Explore Activity

Investigate Energy from the Sun

What happens to the temperature of water when it is in the sun and in the shade?

Science Process Vocabulary

measure verb

When you **measure,** you find out how much or how many.

I can use a thermometer to measure temperature.

compare verb

When you **compare,** you tell how objects or events are alike and different.

The temperature of the hot tea is higher than the temperature of the cold juice.

Materials

2 plastic cups

tape

water

graduated cylinder

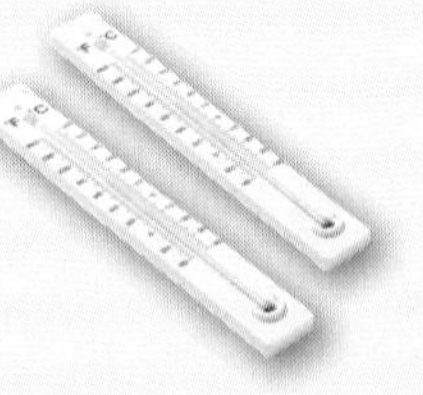

2 thermometers

stopwatch

Next Generation Sunshine State Standards **SC.3.E.6.1** Demonstrate that radiant energy from the Sun can heat objects and when the Sun is not present, heat may be lost. (Also **SC.3.N.1.2**, **SC.3.N.1.3**)

What to Do

1. Label 2 cups **Sun** and **Shade.**

2. Use the graduated cylinder to **measure** 75 mL of water. Carefully pour the water into the Sun cup. Repeat with the Shade cup.

3. Place a thermometer in each cup. Use a stopwatch to time 1 minute. Then measure the temperature of the water in the cups. Record your **data** in your science notebook.

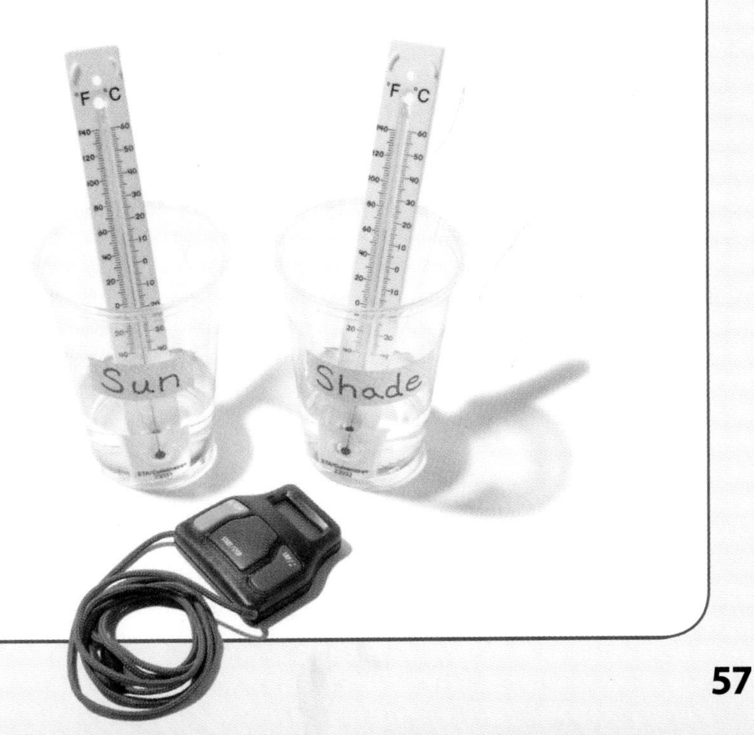

What to Do, continued

4. Place the Sun cup in bright sunlight. Place the Shade cup in the shade. Wait 1 hour. Measure the temperature of the water in each cup again. Record your data.

5. Move the Sun cup to the shady place. Wait 1 hour. Then measure the temperature of the water in both cups. Record your data.

Record

Write in your science notebook.
Use a table like this one.

Water Temperature

Cup	At Start (°C)	After 1 Hour (°C)	After 2 Hours (°C)
Sun		In sun	In shade
Shade		In shade	In shade

Explain and Conclude

1. **Compare** the temperature of the water in the 2 cups at the end of 1 hour in step 4.
2. What happened to the temperature of water in the Sun cup when it was moved from the sunlight to the shade? Explain why you think this happened.
3. **Share** your **data** with others. Explain any differences in the data.

Will the drinks be warmer or colder after sitting in sunlight?

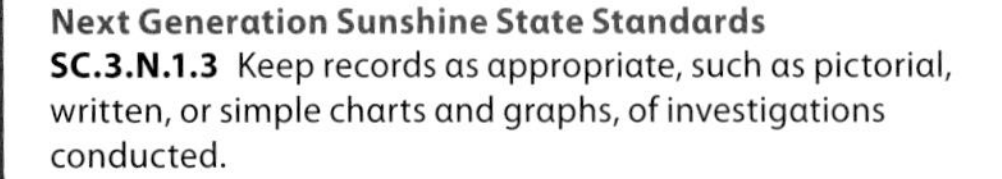

Next Generation Sunshine State Standards
SC.3.N.1.3 Keep records as appropriate, such as pictorial, written, or simple charts and graphs, of investigations conducted.

Math in Science

Bar Graphs

Scientists use a bar graph to show data that is not changing over time. A bar graph has a title that tells what the graph is about. You can tell by the title that the graph is about the distance different model rockets traveled.

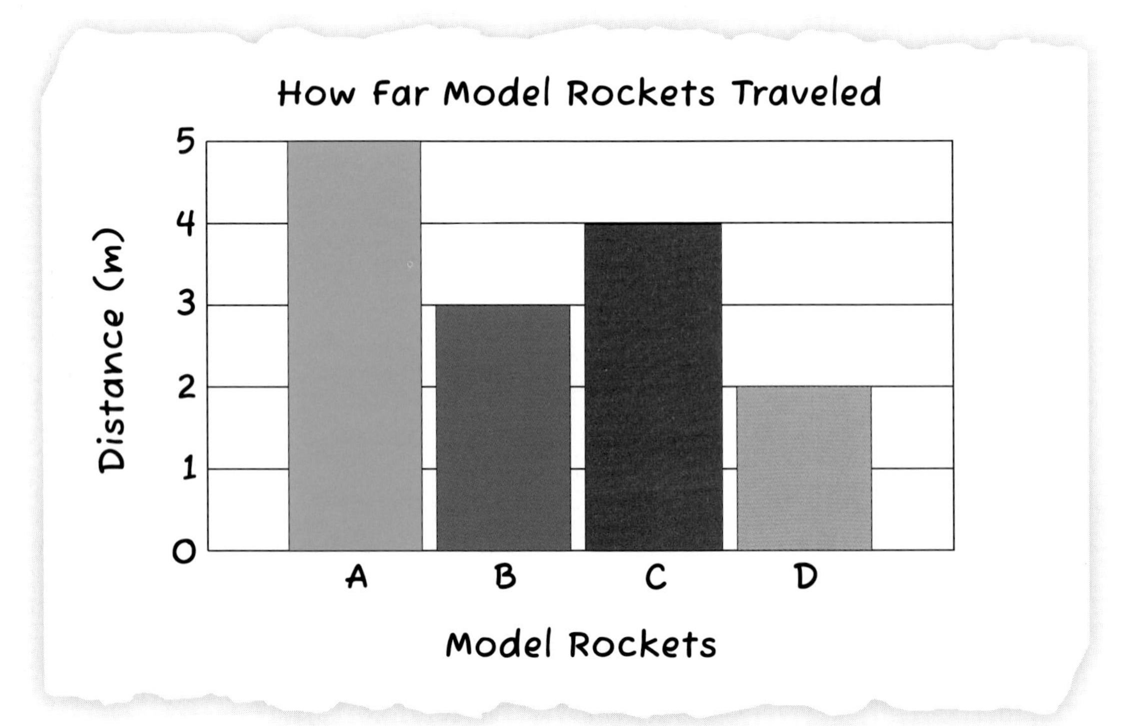

The labels across the bottom of the graph identify the different rockets as A, B, C, or D. A colored bar stands for each rocket.

The rocket that traveled the farthest has the tallest bar. The one that traveled the least distance has the shortest bar.

You can also tell exactly how far each rocket traveled. Look at the numbers along the left side of the graph. This scale tells how many meters were traveled. You know that the numbers stand for meters because the (m) at the end of the label Distance tells you that. You can see that Rocket A traveled 5 m. Rocket D only traveled 2 m.

Making a Bar Graph

You can use a bar graph to organize information you collect during science investigations. Follow these steps.

1. Decide what data you will show. You might want to show how far, how many, or how much of something.

2. Write a title for your graph.

3. Write numbers on the side of the graph, starting at the bottom. Write a label to tell what the numbers mean.

4. Label the bottom of the graph to tell what the different bars stand for.

5. Draw and label your bars. Make each bar a different color.

SUMMARIZE

What Did You Find Out?

1. What kind of data would scientists show on a bar graph?

2. How can you tell what the different bars on a graph stand for?

Make and Use a Graph

Look at one student's data below about the rocks in a particular area. Then use the data to make a graph.

Kind of Rock	How Many Collected
Granite	1
Shale	3
Sandstone	5
Pumice	2

1. Write a title for your graph.
2. Write numbers on the side of the graph and write **How Many Collected.**
3. Write **Kind of Rock** on the bottom of the graph.
4. Write the name of each rock along the bottom. Draw a bar to show how many of each rock were collected.

Share your graph with a partner. Ask and answer questions about your graphs.

Directed Inquiry

Investigate Gravity

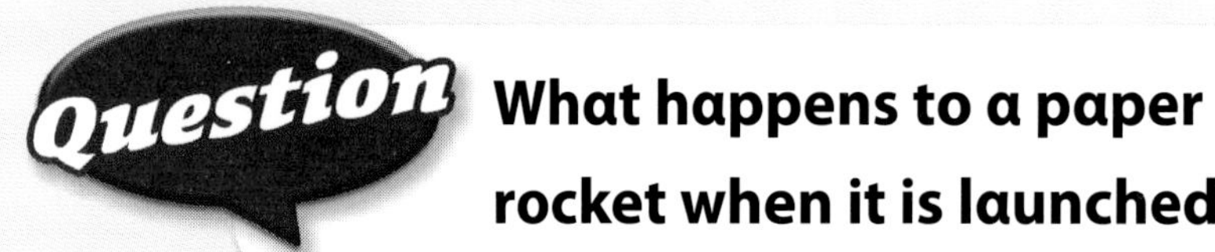

What happens to a paper rocket when it is launched?

Science Process Vocabulary

measure verb

When you **measure,** you find out how much or how many.

I can measure distances with a meterstick.

infer verb

When you use what you know and what you observe to draw a conclusion, you **infer.**

Because the water is falling downward, I infer that gravity is acting on it.

Materials

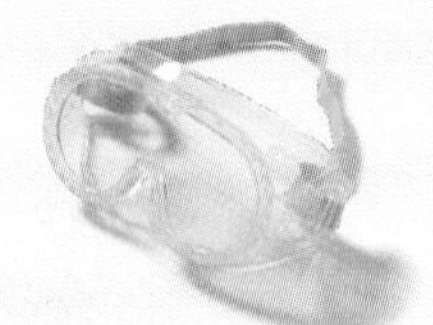
safety goggles

paper circle

scissors

tape

clay

straw

plastic bottle

meterstick

What to Do

1. Put on your safety goggles. Make a cut from the edge of the paper circle to its center. Shape the circle into a cone shape and tape it. Set the cone aside.

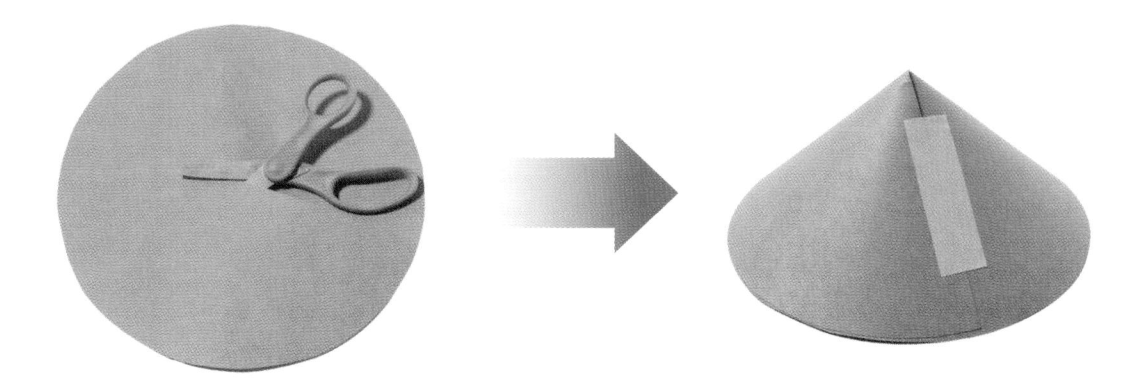

2. Form the clay into a thin rectangle about 2.5 cm long and 1.5 cm wide. Roll the clay around the end of the straw.

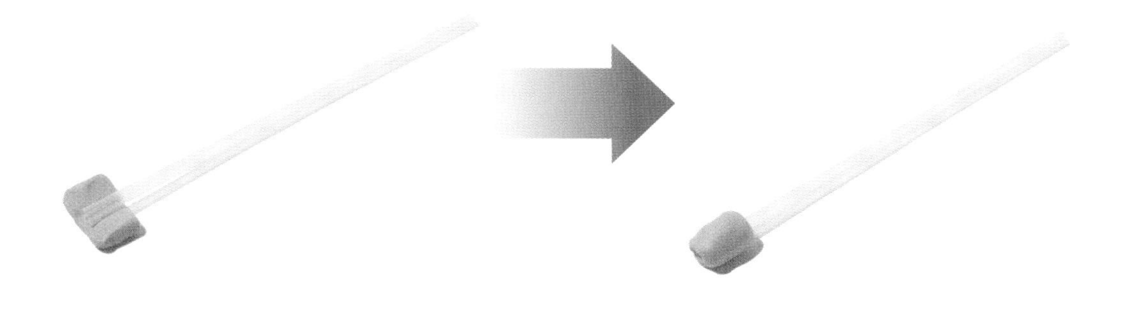

What to Do, continued

3. Push the clay end of the straw inside the top of the bottle. Wrap more clay around the top of the bottle to make the seal tight.

The clay and straw are tightly sealed.

4. Place the paper cone on top of the straw.

5. Make sure the bottle is standing straight. Squeeze the bottle to launch the paper cone "rocket." Use the meterstick to **measure** and record how high the rocket travels.

6. Repeat steps 4 and 5, launching the cone rocket by squeezing the bottle harder. Then repeat steps 4 and 5, squeezing the bottle even harder.

7. Graph your **data.**

Record

Write and draw in your science notebook.
Use a table like this one.

Rocket Motion

How I Squeezed	How High Rocket Traveled (cm)
Hard	
Harder	
Hardest	

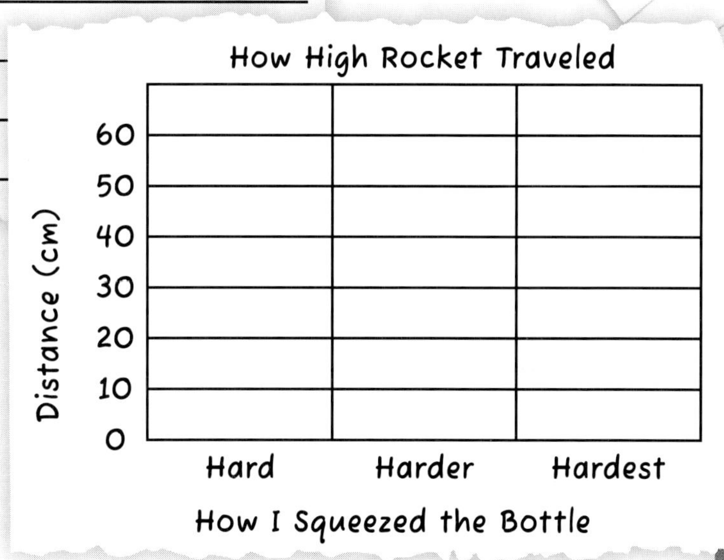

Explain and Conclude

1. How did the distance the rocket traveled change as you squeezed the bottle with different amounts of force?
2. Use your **observations** to **infer** how gravity affected the rockets. Which part of the rocket's path showed that the force of gravity can be overcome?

Think of Another Question

What else would you like to find out about what happens to a paper rocket when it is launched?

Cape Canaveral, Florida

Directed Inquiry

Investigate Light Brightness

How does a light's brightness appear to change with distance?

Science Process Vocabulary

predict verb

When you **predict,** you tell what you think will happen.

I predict that the flashlight will not look as bright if I turn on the lights.

conclude verb

You **conclude** when you use information or data from an investigation to come up with a decision or answer.

I can use my results to conclude which light appears brightest.

Materials

3 penlights

meterstick

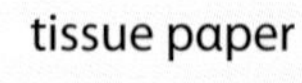

tissue paper

What to Do

1. Label the penlights **A, B,** and **C.** Penlights A, B, and C represent stars that are similar in size and temperature to the sun. Have 3 partners **measure** 2 m away from you. Have them stand at that distance, each holding a light. Have each partner cover the light with tissue paper. The lights should be pointing toward you but should not be turned on.

2. **Predict** how bright each light will look when they turn on their penlights. Record your predictions in your science notebook. Have your partners turn on the lights. **Observe** how bright they appear. Record your observations. Use the Apparent Brightness Scale to describe brightness.

Apparent Brightness Scale	
1	very bright
2	bright
3	dim

What to Do, continued

3. For trial 1, have your partners measure and stand at the following distances. Penlight A should be 2 m away from you. Penlight B should be 4 m away from you. Penlight C should be 3 m away. Then repeat step 2.

4. For trial 2, have your partners measure and stand at the following distances. Penlight A should be 2 m away from you. Penlight B should be 3 m away from you. Penlight C should be 4 m away. Then repeat step 2.

5. For trial 3, change the positions of the lights one more time. Penlight A should be 2 m away. Penlight B should be 5 m away. Penlight C should be 4 m away. Then repeat step 2.

Record

Write in your science notebook.
Use a table like this one.

Distance and Apparent Brightness of Lights

	Penlight	Distance from Observer	Predicted Brightness	Observed Brightness
Start	A	2 m		
	B	2 m		
	C	2 m		
Trial 1	A	2 m		
	B	4 m		
	C	3 m		

Explain and Conclude

1. Do your results support your **predictions?** Explain.
2. What can you **conclude** about distance and the apparent brightness of lights that are the same size? Use your **observations** to support your conclusion.
3. Use the results of this investigation to explain how stars that are like each other in size and temperature can appear to have different brightnesses.

Think of Another Question

What else would you like to find out about how a light's brightness appears to change with distance? How could you find an answer to this new question?

Guided Inquiry

Investigate Lenses

How can lenses help you see objects that are far away?

Science Process Vocabulary

observe verb

When you **observe,** you use your senses to learn about an object or event.

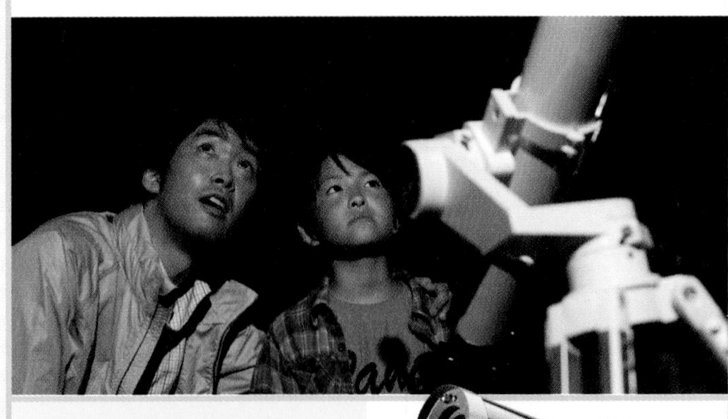

compare verb

When you **compare,** you tell how objects or events are alike and different.

One telescope is older, but they both make the stars look larger.

Materials

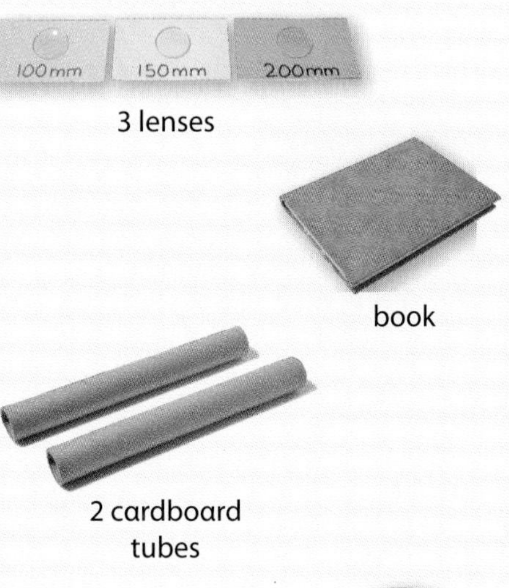

3 lenses

book

2 cardboard tubes

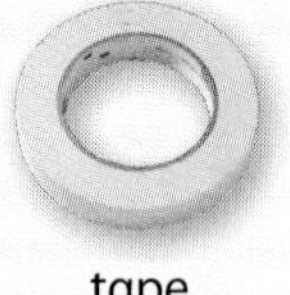

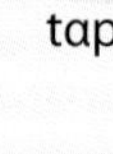

tape

Star Photo

meterstick

Next Generation Sunshine State Standards **SC.3.E.5.5** Investigate that the number of stars that can be seen through telescopes is dramatically greater than those seen by the unaided eye. (Also **SC.3.N.1.1, SC.3.N.1.3**)

What to Do

1. **Observe** the words in the book with all 3 lenses. **Compare** how large the words look with each lens. Record your observations in your science notebook.

2. Choose 2 of the lenses to make a telescope and record your choice. Hold 1 lens carefully in the center on the end of the tube and tape the lens in place. Be careful that the lens does not fall through the tube. Repeat with the other lens and tube. Then slide the tubes together. Make sure the lenses are facing out.

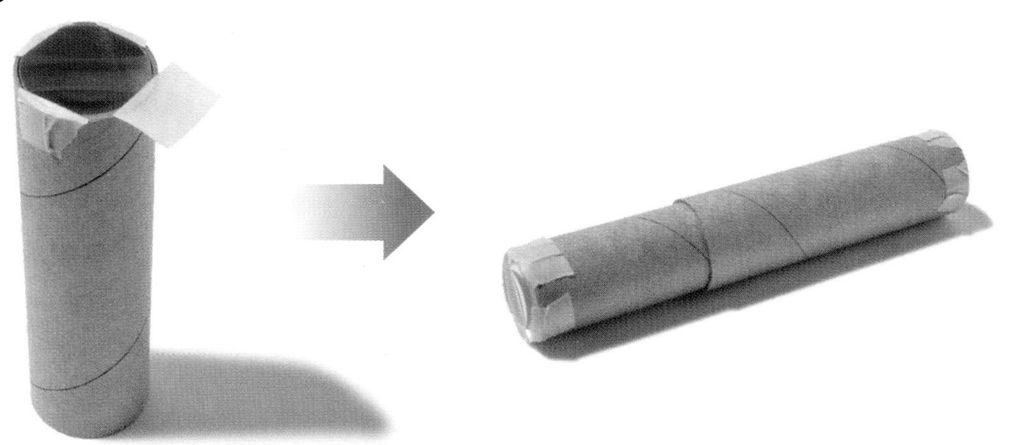

What to Do, continued

3. Tape the Star Photo to a wall. **Measure** 3 m from the photo. Stand at that place. Observe the stars in the picture. Record your observations.

4. Use your telescope to observe the stars. Slide the cardboard tubes in and out until you see a clear image. Record your observations.

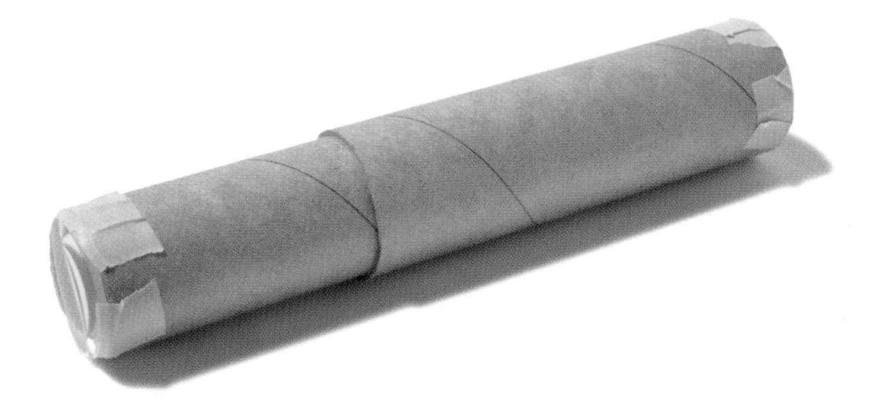

5. Turn the tube around so that the lenses are reversed. Observe the stars. Record your observations.

6. Exchange your telescope with another group. Repeat step 4. Compare what you see through the second telescope with what you saw through your telescope.

Record

Write in your science notebook.
Use a table like this one.

Star Observations

	Observations
Without telescope	
With telescope	
With telescope reversed	

Explain and Conclude

1. How did turning the telescope around affect the stars you **observed?**
2. **Compare** the number of stars you could observe with your telescope and without a telescope.
3. Use your observations to explain how lenses can help scientists study stars.

Think of Another Question

What else would you like to find out about how lenses can help you see objects that are far away? How could you find an answer to this new question?

Scientists use telescopes to study objects in space, such as Comet Hyakutake.

Do Your Own Investigation

Choose one of these questions, or make up one of your own to do your investigation.

- Does light make some colors heat faster than others?
- What can you do to change how far a paper rocket travels?
- What happens to the brightness and color of light if you keep a flashlight on until the batteries stop working?
- How do the colors in sunlight compare to colors in light from a flashlight?

Science Process Vocabulary

First I will find the temperature of the water. Then I will place the water in sunlight.

plan noun

When you make a **plan** to answer a question, you list the materials and steps you need to take.

Open Inquiry Checklist

Here is a checklist you can use when you investigate.

Next Generation Sunshine State Standards
SC.3.N.1.1 Raise questions about the natural world, investigate them individually and in teams through free exploration and systematic investigations, and generate appropriate explanations based on those explorations. (Also **SC.3.N.1.3**)

- ☐ Choose a **question** or make up one of your own.
- ☐ Gather the materials you will use.
- ☐ If needed, make a **hypothesis** or a **prediction.**
- ☐ If needed, identify, manipulate, and control **variables.**
- ☐ Make a **plan** for your **investigation.**
- ☐ Carry out your **plan.**
- ☐ Collect and record **data. Analyze** your data.
- ☐ Explain and **share** your results.
- ☐ Tell what you **conclude.**
- ☐ Think of another question.

Next Generation Sunshine State Standards
SC.3.N.1.3 Keep records as appropriate, such as pictorial, written, or simple charts and graphs, of investigations conducted. (Also **SC.3.N.1.1**)

Write Like a Scientist

Write About an Investigation

Colors and Sunlight

After Sharee stood outside on a cold day, she began to wonder about colors and sunlight. She asked herself, "Will I feel warm faster if I wear certain colors?" She decided to do an investigation. Here is list of what she thought about:

- Sharee decided that using clothing to find the answer would be too difficult.
- She thought she could measure how fast materials heat by making cones out of different colors of paper. She would place the cones over ice cubes to see which ice cube melted fastest.
- She didn't know if the sun would be shining when she did her investigation so she decided to use a lamp to represent the sun.

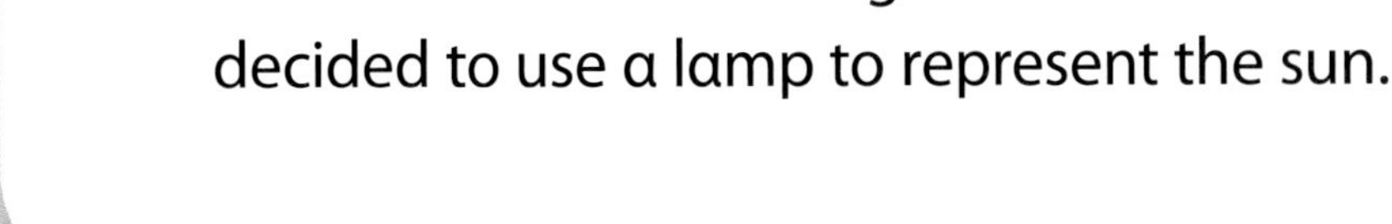

Model

Question

Does light make some colors heat faster than others?

Materials

newspaper

drawing compass

4 sheets of construction paper: white,
black, yellow, and blue

scissors

tape

4 ice cubes

desk lamp

stopwatch

Choose a question that can be answered using materials that are safe and easy for you to get.

List all the materials you will need.

Your Investigation

Now it's your turn to do your investigation and write about it. Write about the following checklist items in your science notebook.

- [ ] **Choose a question or make up one of your own.**
- [ ] **Gather the materials you will use.**

Model

My Hypothesis

If I cover 4 ice cubes with paper cones of different colors, then some ice cubes will melt faster than others. The ice cubes that melt faster will have heated faster.

You can use "If…, then…." statements to make your hypothesis clear.

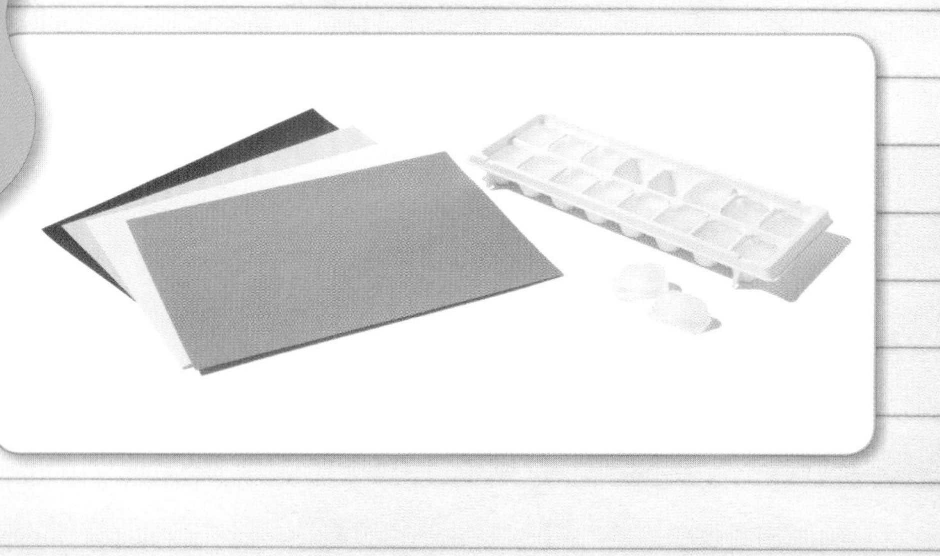

Your Investigation

☐ **If needed, make a hypothesis or prediction.**

Write your hypothesis or prediction in your science notebook.

Model

Variable I Will Change

I will change the color of the paper cone over the ice cube. One ice cube will be covered by white paper, one by black paper, one by yellow paper, and one by blue paper.

During an investigation, it is important to accurately record observations.

Variable I Will Observe or Measure

I will measure how much ice is left for each cube after 20 minutes.

Variables I Will Keep the Same

Everything else will be the same. All the ice cubes and the paper cones will be the same size. All the ice cubes will be the same distance from the light.

Answer these three questions:
1. What one thing will I change?
2. What will I observe or measure?
3. What things will I keep the same?

Your Investigation

☐ **If needed, identify, manipulate, and control variables.**

Write about the variables for your investigation.

Model

My Plan

1. Spread newspaper on a table to keep the area dry.
2. Use the drawing compass to draw a 15 cm circle on each sheet of construction paper. Cut out the circles.
3. Make a cut from the edge of each circle to the center of the circle. Shape each circle into a cone. Tape the edges.
4. Place 4 ice cubes on the newspaper under a lamp.
5. Cover each cube with a different color cone.
6. Turn on the lamp. Observe the ice cubes every 5 minutes for 20 minutes to see how much they have melted.

Write detailed plans. Another student should be able to do your investigation without asking any questions.

Your Investigation

☐ **Make a plan for your investigation.**

Write the steps for your plan.

Model

I carried out all six steps of my plan.

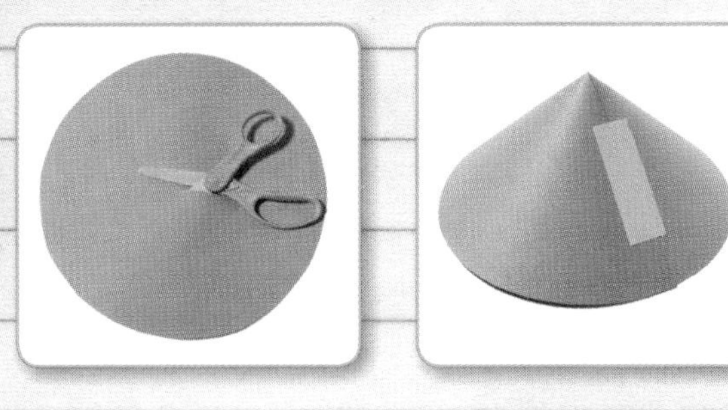

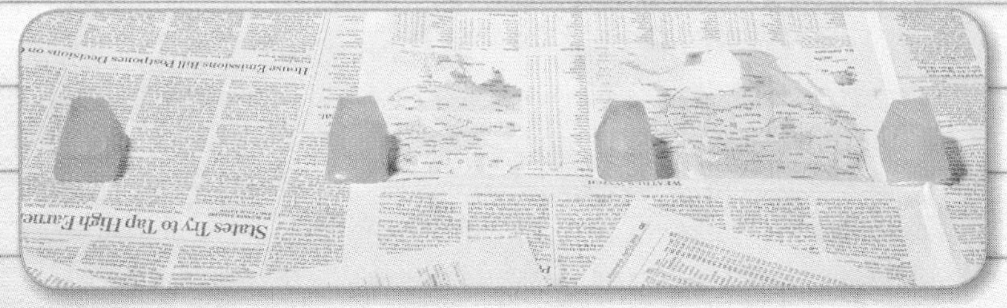

Your Investigation

☐ **Carry out your plan.**

Be sure to follow your plan carefully.

Make a note if you need to change your plan in any way. Sharee did not have to make any changes.

Model

Data (My Observations)

Observations of Ice Cubes

	Amount of Ice Left			
Paper Color	After 5 Minutes	After 10 Minutes	After 15 Minutes	After 20 Minutes
White				
Black				
Yellow				
Blue				

Use a table to organize your data.

My Analysis

At the start, all the ice cubes were the same size.
After 20 minutes, the ice cube under the black cone had melted more than the other ice cubes.

Explain what happened based on the data you collected.

Your Investigation

☐ **Collect and record data. Analyze your data.**

Collect and record your data, and then write your analysis.

Model

How I Shared My Results

First, I explained that I measured how fast ice melts as a way to find out how fast different colors heat up. Next, I drew pictures to show how much ice was left under each color cone. Then, I shared my data.

Scientists often share results so others can find out what was learned.

My Conclusion

The ice cube under the black cone melted faster than the ice cube under the other cones. That meant that the black construction paper heated up faster than the other construction paper. The results of the investigation support my hypothesis.

Tell what you conclude and what evidence you have for your conclusion.

Another Question

I wonder if some surfaces heat faster in sunlight. Do dull surfaces heat faster than shiny surfaces?

Investigations often lead to new questions for inquiry.

my SCIENCE notebook

Your Investigation

- ☐ Explain and share your results.
- ☐ Tell what you conclude.
- ☐ Think of another question.

Next Generation Sunshine State Standards
SC.3.E.6.1 Demonstrate that radiant energy from the Sun can heat objects and when the Sun is not present, heat may be lost.

Science and Technology

Using Solar Energy

Solar energy is energy from the sun. It is Earth's most plentiful energy source. Solar energy is a renewable energy source. A renewable energy source is a resource that will not run out.

Energy from the sun can be changed into other forms of energy, such as heat and electricity.

Solar energy can be changed into electricity using a solar cell. Most calculators contain a small solar cell.

The small, flat solar cell can change solar energy into electricity to run a calculator.

Solar Power Plants Some sunny, warm states are developing solar energy power plants. These power plants use solar energy to make electricity.

This solar farm in Arcadia, Florida, can make electric power for many homes.

Solar Water Heating Solar energy is also used to heat water in people's homes. The simplest solar water heating systems are called passive solar water heaters. This type of solar heater is most common in places without long periods of freezing temperatures, such as in much of Florida.

Solar energy collector (water is heated)

Cold water in

Hot water out

Tank to store hot water

Passive solar water heaters are made of two main parts: a solar collector and a tank to store the heated water.

People use a solar cooker to cook rice in a village in Zambia, Africa.

Solar Ovens You may know that sunlight can make objects warm, or even hot. Hundreds of years ago, people used sunlight to start fires. Then about 200 years ago, a scientist built a solar oven.

Today different kinds of solar ovens and solar cookers are used all over the world. Solar cookers are cheap to make and easy to use. People can easily carry them from one place to another. Most importantly, solar cookers can heat food and water hot enough to make them safe to eat and drink.

SUMMARIZE

What Did You Find Out?

1. What is solar energy?
2. Why is solar energy a good source of energy?
3. How is solar energy used?

 Observe Solar Energy

You can observe how the sun's energy heats objects. Follow the steps below using 3 smooth black rocks.

1. Build a "house" for 1 of the rocks. Place a small box inside a larger box. Use paper towels to fill the spaces between the boxes.

2. Put the rock inside the house and cover the opening with plastic wrap and a rubber band. Place the house in a sunny spot so that the sunlight shines directly on the rock.

3. Place the other 2 rocks in the sunlight next to the house. Cover 1 rock with a clear plastic cup.

4. After 30 minutes, feel each rock. Which rock feels the warmest? Explain why you think this is so.

Snap!

Next Generation Sunshine State Standards **SC.3.P.8.2** Measure and compare the mass and volume of solids and liquids.

CHAPTER 6

Science in a Snap! Water Predictions

Observe the 2 containers. **Predict** which container has the most water. Then pour the water from 1 container into a measuring cup. **Measure** and record the amount of water in the cup. Pour the water back into its container. Repeat for the other container. Was your prediction correct? Explain.

Next Generation Sunshine State Standards **SC.3.P.9.1** Describe the changes water undergoes when it changes state through heating and cooling by using familiar scientific terms, such as melting, freezing, boiling, evaporation, and condensation. (Also **SC.3.N.1.6**)

Science in a Snap! Observe Water Drops

Mix a few drops of food coloring into a metal can of ice water. Cover the can with plastic wrap. **Observe** the drops of water that form on the outside of the can. Wipe the drops of water with tissue paper. **Compare** the color of the drops to the water inside the can. Where do you think the drops of water came from?

Next Generation Sunshine State Standards **SC.3.P.10.1** Identify some basic forms of energy such as light, heat, sound, electrical, and mechanical.

CHAPTER 8

Science in a Snap! Observe Sound

Put some water in a plastic bag and seal the bag tightly. Put the bag over one ear. Have a partner lightly tap on the bag. **Observe** how loud the tapping seems. Repeat with a bag filled with air. **Compare** how well you hear the tapping. How does tapping the bag cause the sounds you hear?

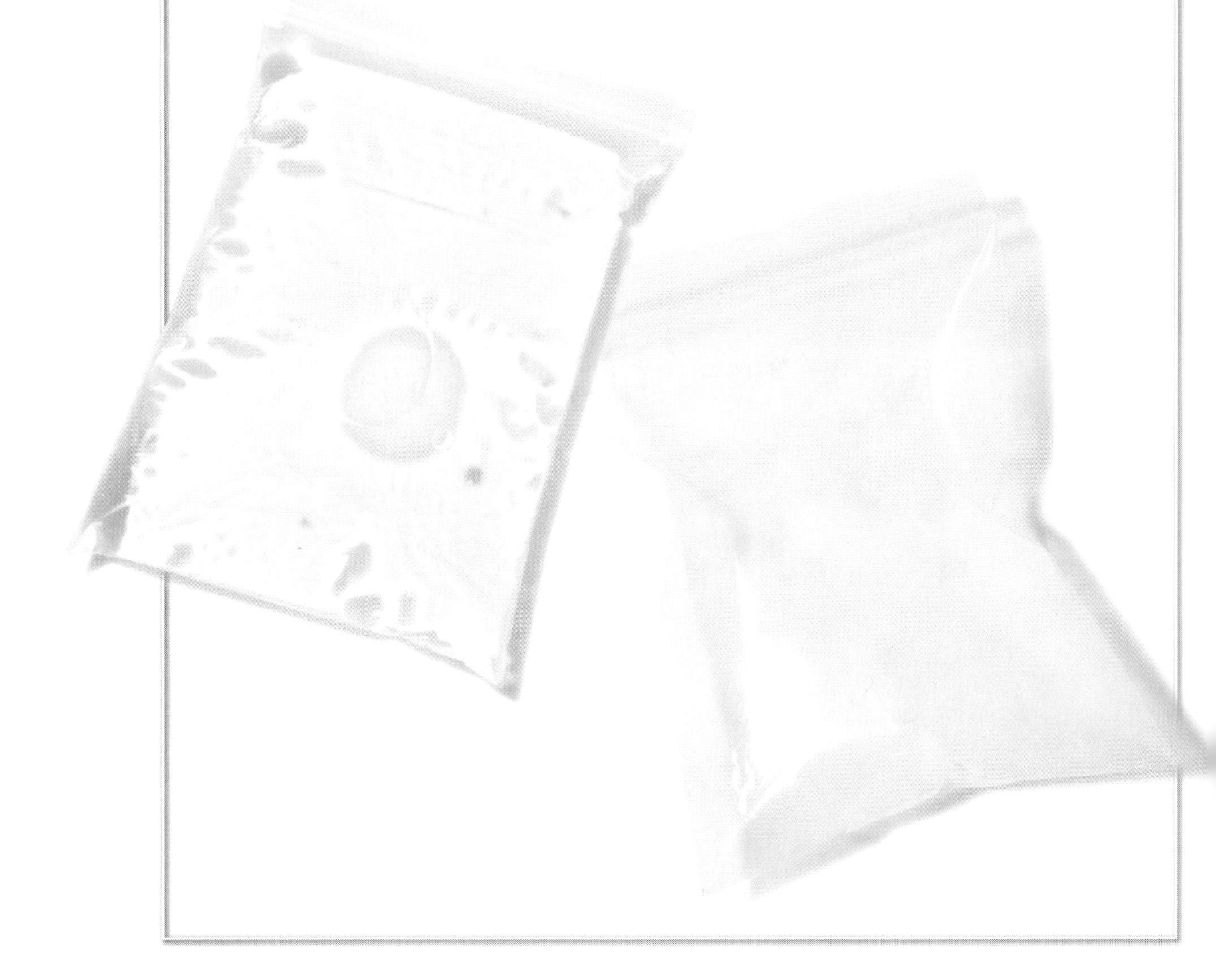

Science in a Snap! Observe Heat

Cut out squares of white construction paper, black construction paper, foil, tissue, and plastic wrap. Line up the materials in front of a lamp. Turn on the lamp. Wait 10 minutes, then turn off the lamp. **Compare** how warm each material feels. Why might some materials feel warmer than others?

Explore Activity

Investigate Properties of Objects

How can you classify objects by their properties?

Science Process Vocabulary

classify verb

When you **classify,** you put things in groups by their characteristics.

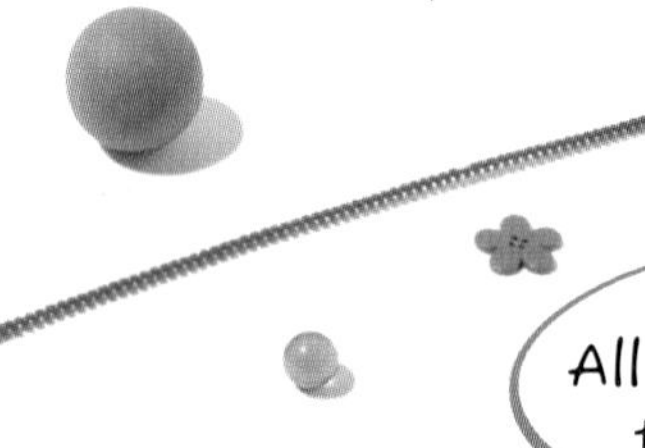

compare verb

You **compare** when you find out how things are alike and how they are different.

Materials

rock

marble

craft stick

chenille stem

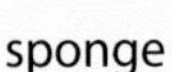
sponge

eraser

ruler

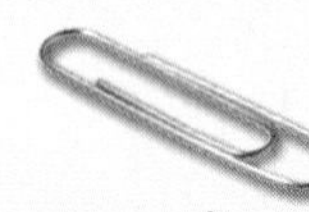
paper clip

hand lens

What to Do

1. **Observe** the shape of each object. Draw each object in your science notebook. Record the color of each object.

2. Use the metric ruler to **measure** the length of each object. Record your **data.**

3. Texture is how an object feels. Rub your finger over each object to observe its texture. Record your observations.

What to Do, continued

4. Test how hard the objects are. Tap each one with a paper clip. Try to bend each object. Record your observations.

5. Use the hand lens to observe each object. Record your observations. Tell what you can see with the hand lens that you could not see without it.

6. Choose one property you observed to **classify** the objects into 2 groups. Record the property and the objects in each group. Then sort the objects in each group by another property of your choice. Record the property and the objects.

Record

Write and draw in your science notebook.
Use tables like these.

Properties of Objects

Object	Shape	Color	Length (cm)	Texture	Hardness	With Hand Lens
Rock						

Classify Objects

Property	Objects in Group 1	Objects in Group 2

Explain and Conclude

1. What properties did you use to **classify** the objects into two groups? What objects did you place in each group?

2. **Compare** how different groups in your class classified objects. What property was used most often to classify objects?

What properties could you use to classify these balloons?

Directed Inquiry

Investigate Volume and Mass

Question **How can you compare the volume and mass of solid and liquid objects?**

Science Process Vocabulary

measure verb

You can use tools to **measure** mass and volume.

predict verb

You **predict** when you say what you think your results will be in an investigation.

I predict that the marble has more mass than the rock.

Materials

graduated cylinder

water

marble

rock

cup

balance

gram masses

Next Generation Sunshine State Standards SC.3.P.8.2 Measure and compare the mass and volume of solids and liquids. (Also **SC.3.N.1.1**, **SC.3.N.1.2**, **SC.3.N.1.3**)

What to Do

1 Put 20 mL of water into the graduated cylinder. Add the marble. **Measure** the volume of the water and the marble. Record your **data** in your science notebook.

2 To find the volume of the marble, subtract the volume of the water from the volume of the water and the marble. Record your data.

3 Repeat steps 1–2 with the rock.

4 Now you will measure mass. Which do you think has the most mass—the marble, the rock, or 20 mL of water? Write your **prediction.**

What to Do, continued

5. Use the balance to find the mass of the cup. Record your data.

6. Place the marble in the cup and find the mass of the cup and marble. Record your data. To find the mass of the marble, subtract the mass of the cup from the mass of the cup and marble. Record your data.

7. Repeat step 6 with the rock and with 20 mL of water.

Record

Write in your science notebook.
Use tables like these.

Volume

Object	Volume of Water	Volume of Water and Object	Volume of Object
Marble	20 mL		
Rock	20 mL		

Mass

Object	Mass of Cup	Mass of Cup and Object	Mass of Object
Marble			
Rock			
Water			

Explain and Conclude

1. **Compare** the volume of the water, marble, and rock. Which had the most volume?
2. Which object had the most mass? Is that what you **predicted?** Explain.
3. **Share** your results with others. Explain any differences.

Think of Another Question

What else would you like to find out about comparing the volume and mass of solids and liquids? How could you find an answer to this new question?

Early balance

Directed Inquiry

Investigate Water and Temperature

What happens to water as the temperature changes?

Science Process Vocabulary

observe verb

When you **observe,** you use your senses to learn about objects or events.

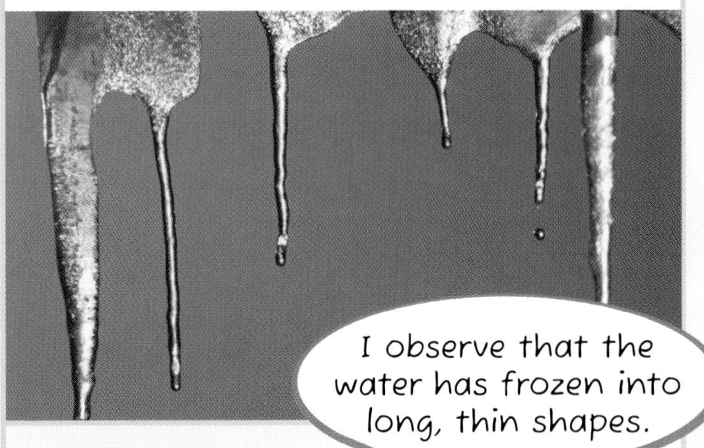

compare verb

When you **compare,** you tell how objects or events are alike and different.

The icicles are all clear and cold, but they have different shapes.

Materials

2 resealable bags

tape

graduated cylinder

water

Next Generation Sunshine State Standards **SC.3.P.9.1** Describe the changes water undergoes when it changes state through heating and cooling by using familiar scientific terms, such as melting, freezing, boiling, evaporation, and condensation. (Also **SC.3.N.1.3, SC.3.N.1.6, SC.3.N.1.7**)

1. Label the plastic bags **Bag 1** and **Bag 2.** Use the graduated cylinder to **measure** 100 mL of water. Pour the water into one bag. Seal the bag. Repeat with the other bag.

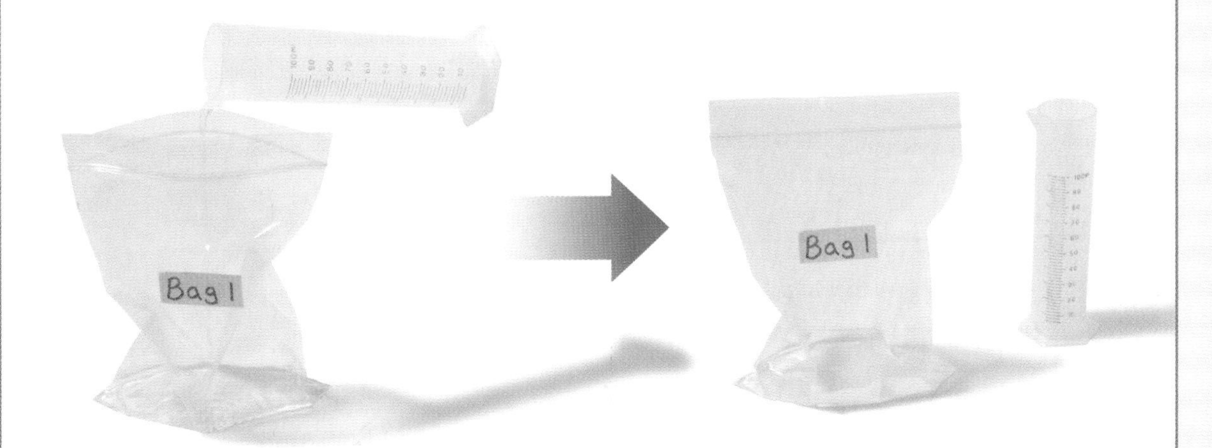

2. Carefully place both bags in the freezer. **Predict** what will happen to the water in the bags. Record your prediction in your science notebook.

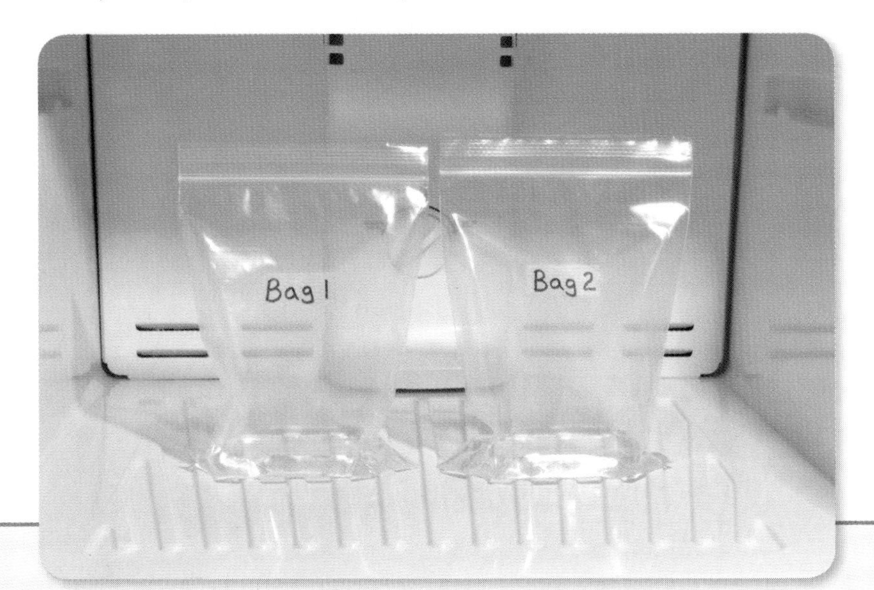

What to Do, continued

3. The next day, take the bags out of the freezer. Turn the bags in different directions. **Observe** the shape of the water. Record your observations.

4. Put the bags on your desk. Wait 30 minutes. Turn the bags in different directions. Record your observations.

5. Put the bags in sunlight. Open Bag 1, being careful not to spill the water. Keep Bag 2 sealed. Predict what will happen to the water in both bags after 3 days.

6. Observe the bags every day for the next 3 days. Record your observations.

Record

Write or draw in your science notebook.
Use a table like this one.

Observations of Bags with Water

	Bag 1	Bag 2
After 1 day in the freezer		
After 30 minutes on the desk		

Explain and Conclude

1. Did your results support your **predictions?** Explain.
2. **Compare** your results with the results of other groups. What patterns do you see?
3. Use your results from steps 2–6 to **conclude** what happens to water when the temperature goes down. What happens to frozen water when the temperature goes up?

Think of Another Question

What else would you like to find out about what happens to water as the temperature changes? How could you find an answer to this new question?

What will happen to the water on this tomato as time passes?

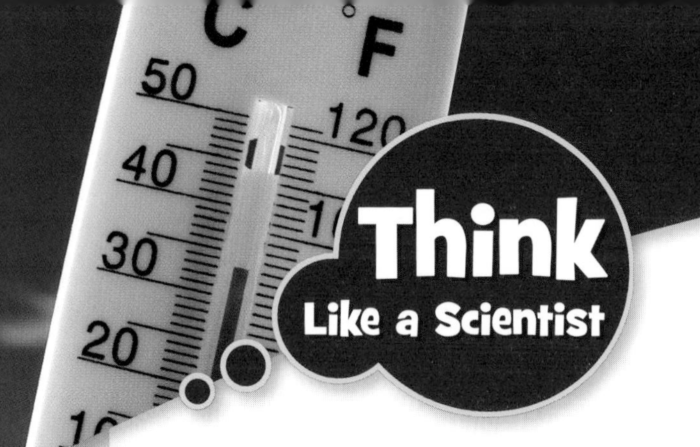

Next Generation Sunshine State Standards
SC.3.N.1.2 Compare the observations made by different groups using the same tools and seek reasons to explain the differences across groups.

Math in Science

Measuring Temperature

Scientists use thermometers to measure temperature. Thermometers help scientists make better observations. Scientists can get a more exact temperature with a thermometer than by just feeling how hot or cold something is. Also, scientists can use thermometers to measure the temperatures of objects that are too hot or too cold to touch.

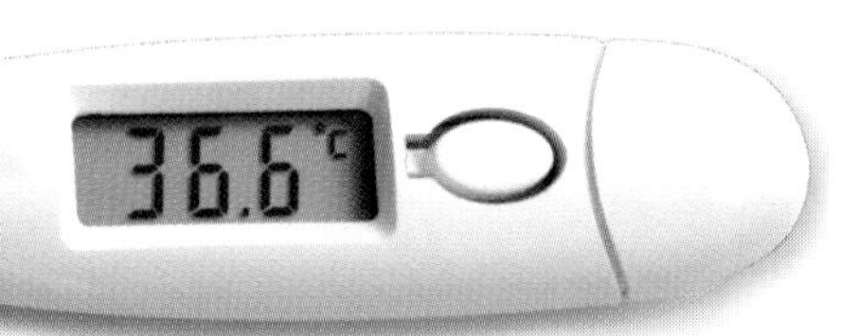

Digital thermometer

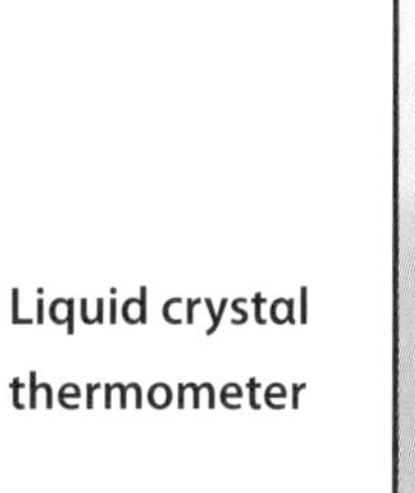

Liquid crystal thermometer

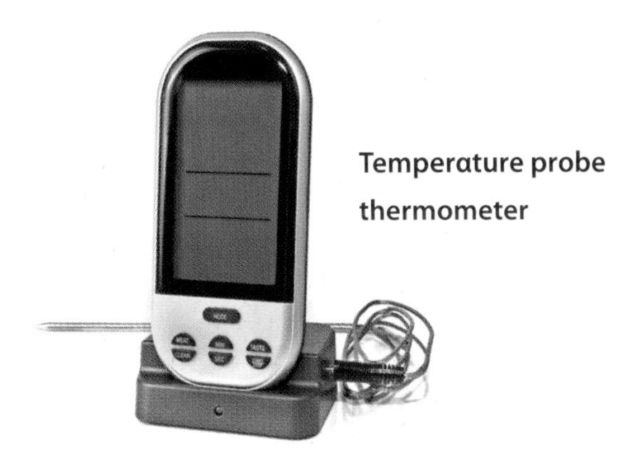

Temperature probe thermometer

The simplest thermometer is made up of a clear tube that contains a liquid. As the temperature gets warmer, the liquid rises in the tube. When the temperature gets cooler, the liquid moves lower in the tube.

Temperature Scales Look at the scale on both sides of the thermometer. One scale shows the temperature in degrees Fahrenheit (°F). You may have heard the temperature given in degrees Fahrenheit during a television weather report.

The other scale shows the temperature in degrees Celsius (°C). Scientists usually measure temperature in degrees Celsius.

Using a Thermometer

When reading a thermometer, follow the steps below. For steps 2 and 3 use the thermometer in the photograph to practice.

1. Place the thermometer on or in the material of which you want to find the temperature. Wait about 1 minute for the thermometer to complete its reading.
2. Put the top of your finger at the top of the red liquid.
3. Slide your finger to the right to read the Celsius temperature. Slide your finger to the left to read the Fahrenheit temperature.

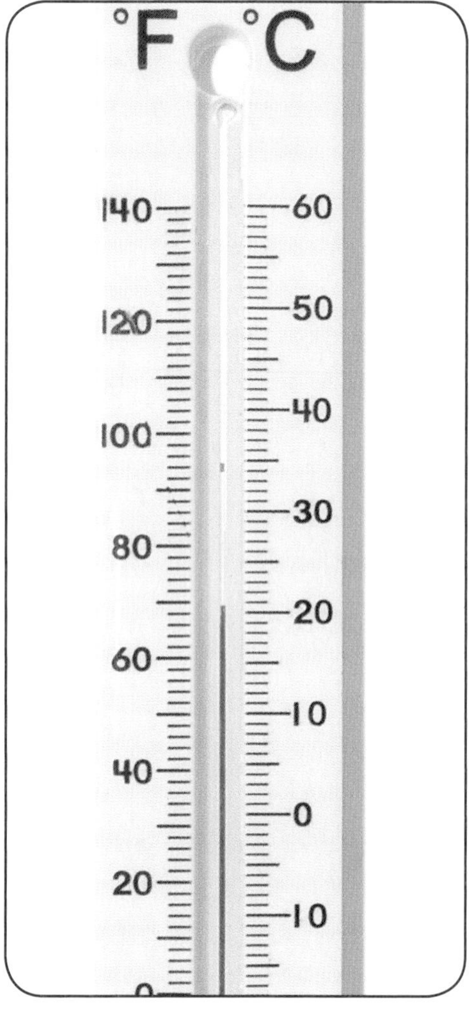

When you measure temperature, make sure you understand the number of degrees each mark shows.

SUMMARIZE

What Did You Find Out?

1. What are two reasons why scientists use thermometers to measure temperature?
2. What two scales are used to measure temperature? Which do scientists usually use?

Measure the Temperature of Water

Practice using a thermometer. Follow these steps:

1. Fill a plastic cup half full of water.
2. Place a thermometer in the water. Wait 1 minute.
3. Find the temperature of the water on both the Fahrenheit and Celsius scales. Record your measurements.
4. Switch cups with a partner. Measure the temperature in your partner's cup of water.
5. Compare your temperature readings with your partner. Explain why there might be differences between your measurement and your partner's measurement.

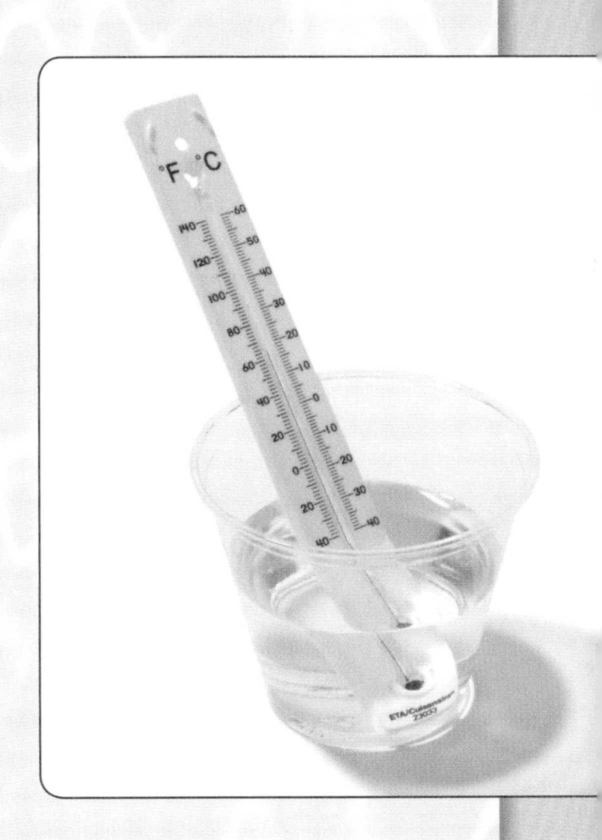

Guided Inquiry

Investigate Temperature

How can you compare how fast water and other materials heat and cool?

Science Process Vocabulary

hypothesis noun

You make a **hypothesis** when you state a possible answer to a question that can be tested by an experiment.

My hypothesis is that on a sunny day the temperature of the sand will go up 15° C.

Materials

safety goggles

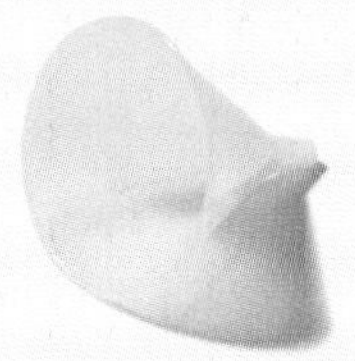
funnel

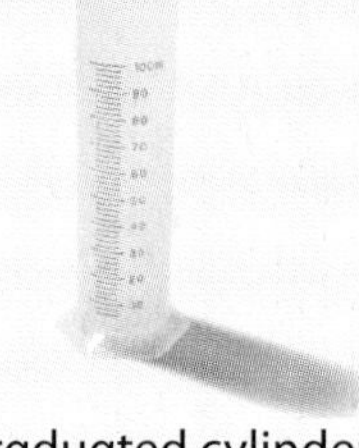
graduated cylinder

sand

salt

2 plastic cups

water

oil

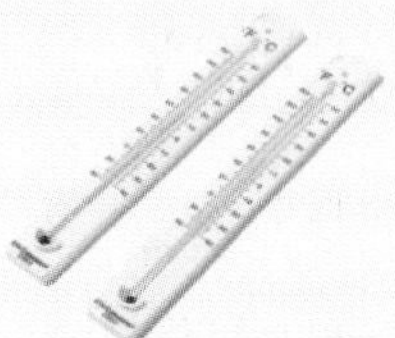
2 thermometers

stopwatch

Do an Experiment

Write your plan in your science notebook.

Make a Hypothesis

In this investigation, you will choose a solid and a liquid to test. Then you will compare how the solid and the liquid heat up and cool down. Which solid will you choose? Which liquid? Which material will heat up and cool down more quickly? Write your **hypothesis.**

Identify, Manipulate, and Control Variables

Which variable will you change?
Which variable will you observe or measure?
Which variables will you keep the same?

What to Do

1. Put on your safety goggles. Select 1 liquid and 1 solid to investigate. Use the funnel to pour 100 mL of the solid into the graduated cylinder. Pour the solid into a plastic cup.

2. **Measure** 100 mL of the liquid. Pour the liquid into a plastic cup.

What to Do, continued

3. Place a thermometer in each cup. Measure the temperature of the liquid and the solid. Record your **data.**

4. Place both cups in sunlight. Record the temperature of the solid and the liquid after 20 minutes.

5. Move the cups to the shade. Record the temperature of the solid and the liquid after 20 minutes.

Record

Write in your science notebook.
Use a table like this one.

Heating and Cooling

	Material Used	Beginning Temperature of Material (°C)	Temperature of Material After 20 Minutes in Sunlight (°C)	Temperature of Material After 20 Minutes in Shade (°C)
Liquid				
Solid				

Explain and Conclude

1. Did your results support your **hypothesis?** Explain.
2. **Compare** your results with the results of other groups. What patterns do you see?

Think of Another Question

What else would you like to find out about how fast different materials heat up and cool down? How could you find an answer to this new question?

Special cameras can take images of heat given off when this child rubs his hands together. The white areas are warmest.

Directed Inquiry

Investigate Energy of Motion

Question How does adding more washers affect the motion of a pendulum?

Science Process Vocabulary

count verb

When you **count,** you tell the number of something.

compare verb

When you **compare,** you tell how objects or events are alike and different.

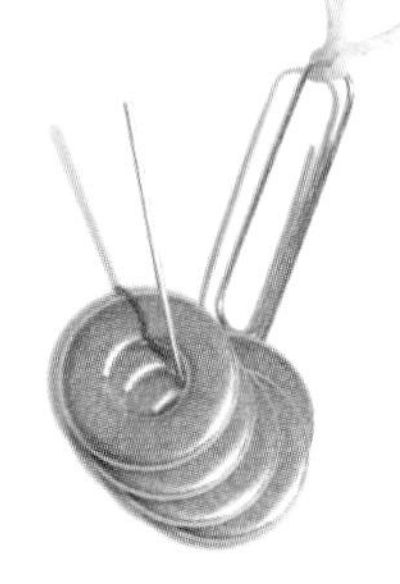

Materials

safety goggles

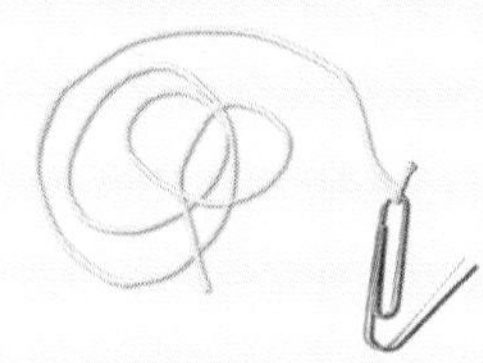

string with paper clip hook

tape

metal washers

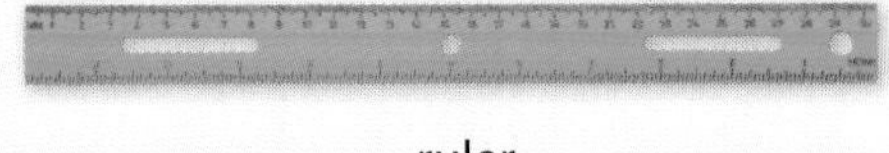

ruler

stopwatch

What to Do

1. Put on your safety goggles. Tape the free end of the string to the edge of a table. Put a metal washer on the paper clip hook. The metal washer should be close to the floor but not touching it. You have made a pendulum.

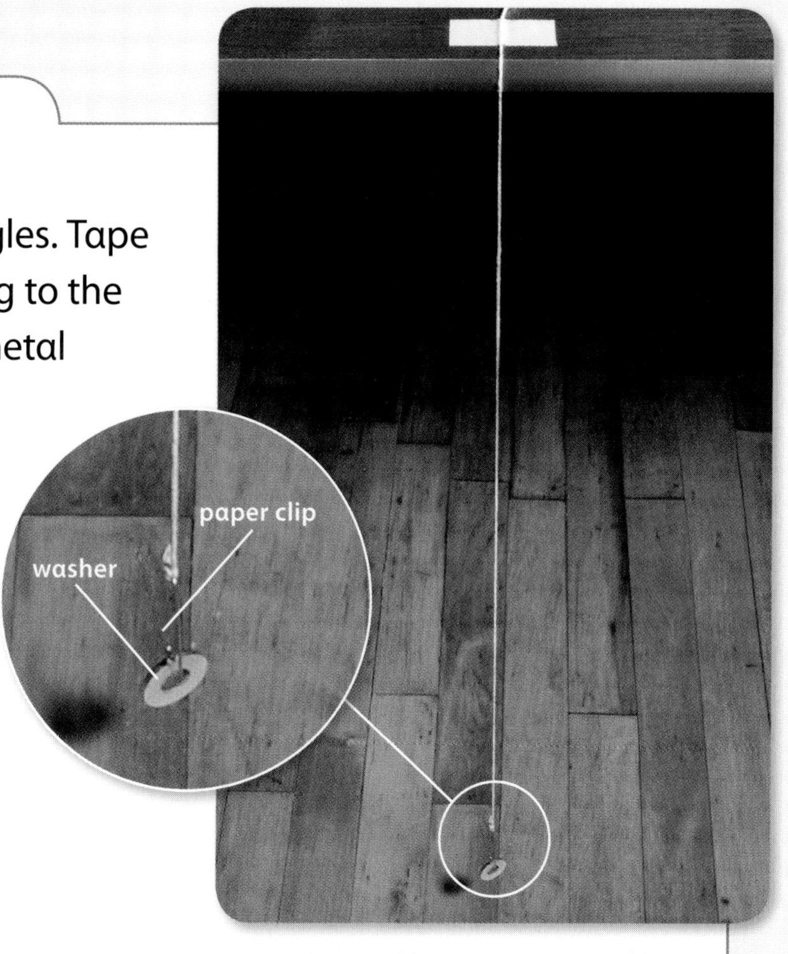

2. Pull the pendulum up 10 cm, and then let it go. Use the stopwatch to **count** the number of swings the pendulum makes in 1 minute. One swing is when the pendulum moves back and forth 1 time. Record your **data** in your science notebook.

What to Do, continued

3. Add 2 more washers to the paper clip hook. How do you think adding more washers will affect the number of pendulum swings in 1 minute? Record your **prediction.** Repeat step 2. Be sure to pull the pendulum up 10 cm before letting it go.

4. Add 2 more washers to the paper clip hook. How do you think adding even more washers will affect the number of pendulum swings in 1 minute? Record your prediction. Repeat step 2.

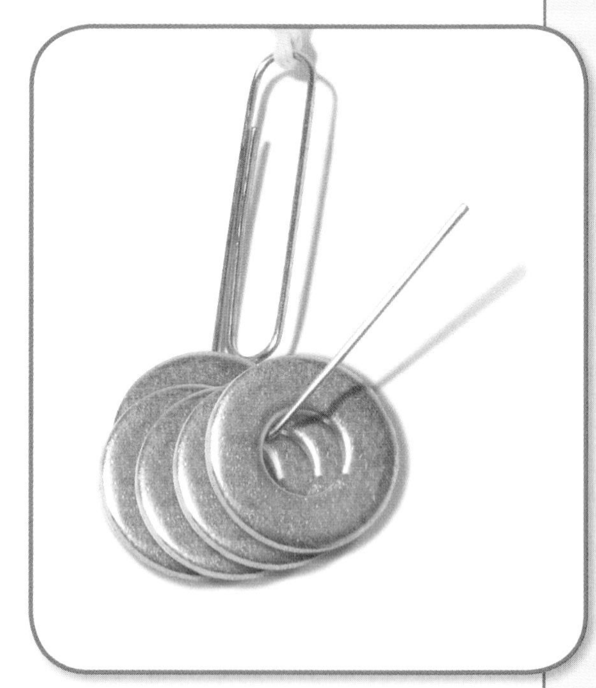

5. Use your data to make a bar graph to **compare** the pendulum swings. Look for a pattern in the graph.

Record

Write in your science notebook. Use your data to make a bar graph like the one shown here.

Pendulum Swings

Number of Washers	Prediction	Number of Swings
1		
3		
5		

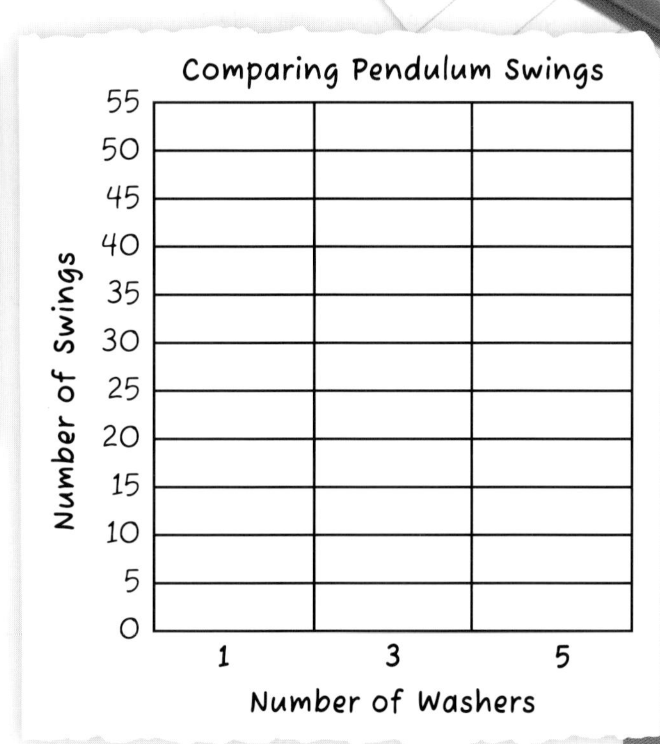

Explain and Conclude

1. Did your results support your **predictions?** Explain.
2. **Compare** the number of swings the pendulum made with 1, 3, and 5 washers. How did increasing the number of washers affect the motion of the pendulum?
3. Describe the motion of the pendulum. Where in its swing does the pendulum move fastest? Where does it move slowest?

Think of Another Question

What else would you like to find out about how adding more washers can affect the swing of a pendulum? How could you find an answer to this new question?

Guided Inquiry

Investigate Motion

How can you use the energy of moving air to move a model car?

Science Process Vocabulary

model noun

You can use a **model** to find out how something works.

You can make a model car out of simple materials.

I can use mechanical energy to make the car move.

Materials

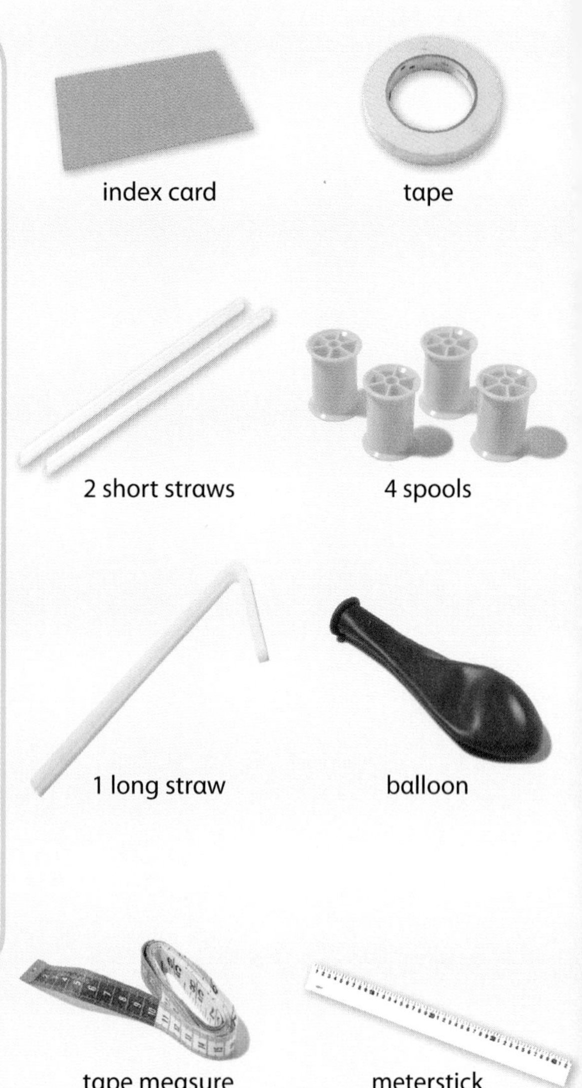

What to Do

1. Fold an index card in half and tape it closed. Tape the short straws to the index card.

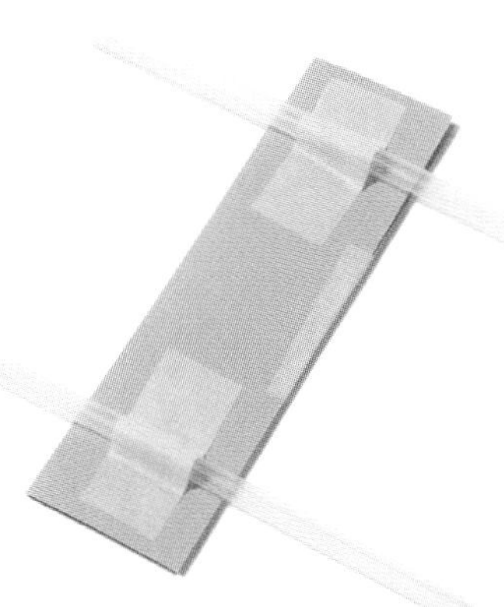

2. Put spools over the ends of the straws. Place a piece of tape on the end of each straw to keep the spools from falling off.

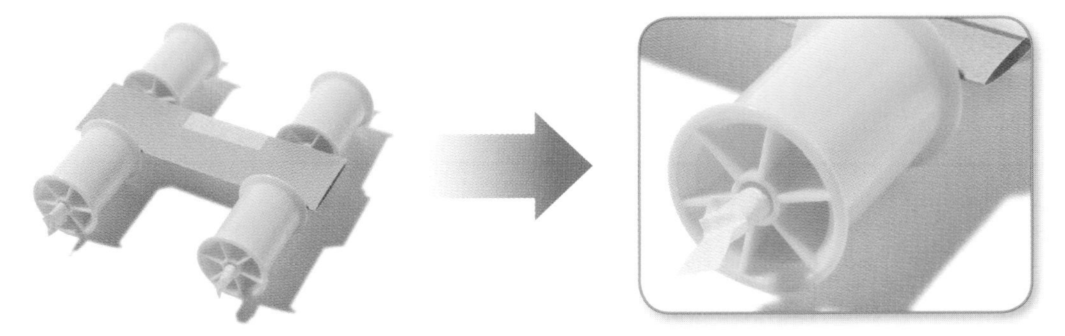

3. Tape the long straw onto the top of the index card. Use tape to attach the balloon to the straw. You have made a **model** car.

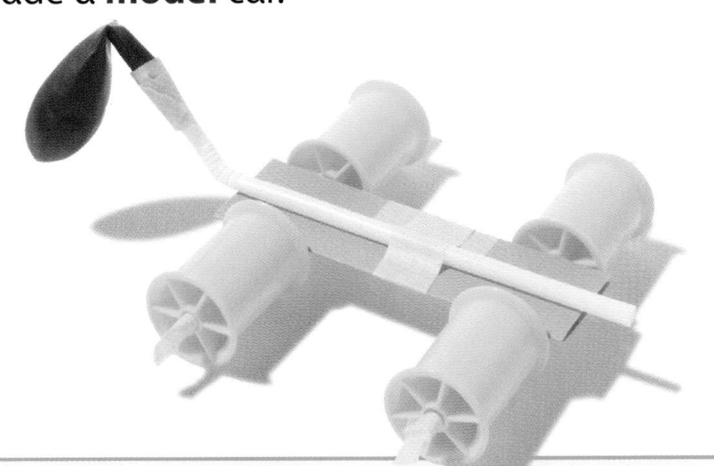

What to Do, continued

4. Mark a **Start Line** on the floor with tape.

5. Blow through the straw on your model car to partially inflate the balloon. Hold a finger over the end of the straw to keep the air from escaping. Have a partner use the tape measure to **measure** the balloon around its widest part. Record your measurement.

6. Place the car on the **Start Line.** Remove your finger from the end of the straw. Use the meterstick to measure how far the car goes. Record your **data.**

7. Repeat steps 5 and 6, but this time blow more air into the balloon. Record your results.

Record

Write in your science notebook.
Use a table like this one.

Travel Distances of Car

	Measurement Around Balloon (cm)	Distance Traveled (cm)
Trial 1 (less air)		
Trial 2 (more air)		

Explain and Conclude

1. Moving air provided the energy to move the car. What kind of energy does moving air have? How were you able to change the amount of energy from moving air?
2. In which trial did the car move farther? Why do you think this happened?
3. **Share** your results with other groups. Explain any differences.

Think of Another Question

What else would you like to find out about how you can use the energy of moving air to move a model car? How could you find an answer to this new question?

Directed Inquiry

Investigate Light and Heat

What happens to an object's temperature when light shines on it?

Science Process Vocabulary

measure verb

When you **measure,** you find out how much or how many.

conclude verb

When you **conclude,** you use information, or data, from an investigation to come up with a decision or answer.

I conclude that the temperature of the sand is higher than the temperature of the water.

Materials

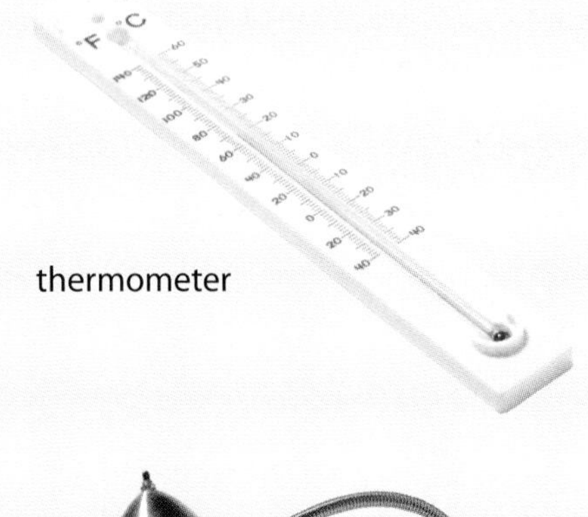

thermometer

lamp

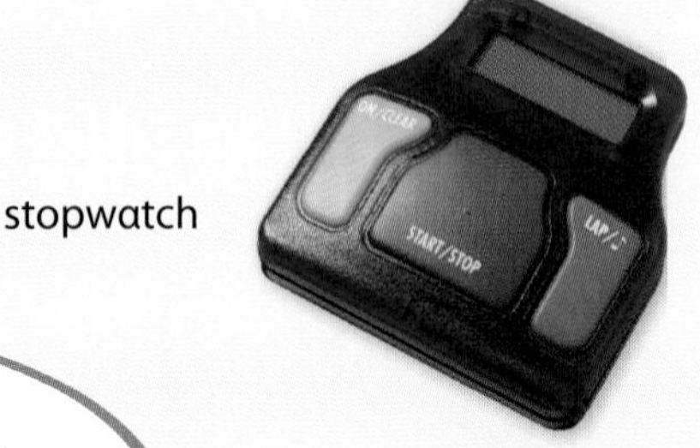

stopwatch

Next Generation Sunshine State Standards **SC.3.P.11.1** Investigate, observe, and explain that things that give off light often also give off heat. (Also **SC.3.N.1.1, SC.3.N.1.3**)

What to Do

1. Put the thermometer on your desk. Place the lamp over the thermometer. Do not turn on the lamp. Wait 5 minutes. Then record the temperature on the thermometer in your science notebook as the Start temperature.

2. **Predict** what will happen to the temperature if you turn on the lamp. Record your prediction.

3. Turn on the lamp. Wait 5 minutes. **Measure** and record the temperature on the thermometer.

What to Do, continued

4. Predict what will happen to the temperature if you leave the lamp on for 5 more minutes. Record your prediction. Then wait 5 minutes. Record the temperature on the thermometer.

5. Predict what will happen to the temperature if you turn the lamp off. Record your prediction. Turn off the lamp. Move the thermometer to a place on your desk that has not been heated by the lamp. Wait 5 minutes. Record the temperature on the thermometer.

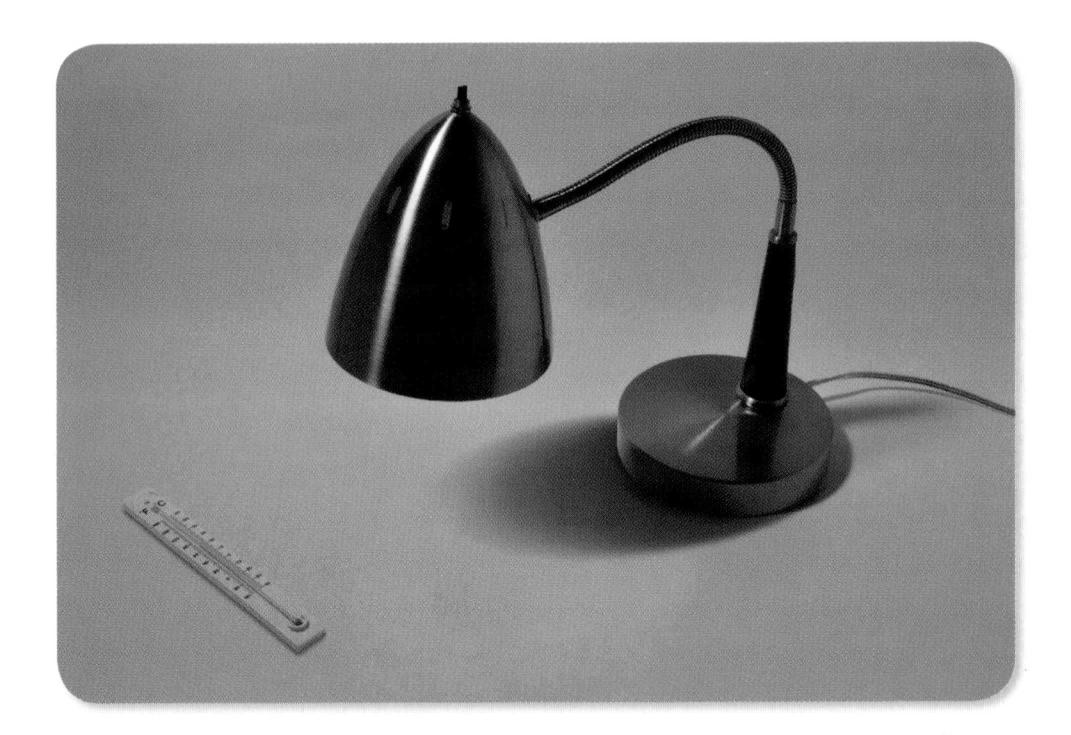

6. Predict what will happen to the temperature if you leave the lamp off for 5 more minutes. Record your prediction. Wait 5 more minutes. Then, record the temperature on the thermometer again.

Record

Write in your science notebook.
Use a table like this one.

Light and Temperature

	Temperature(°C)	Predictions
Start		What will happen to the temperature if you turn on the lamp?

Explain and Conclude

1. Did your **observations** support your **predictions?** Explain.
2. What happened to the temperature as the lamp shined on the thermometer longer? What happened after you turned off the lamp?
3. What can you **conclude** about what can happen to the temperature of an object when light shines on it?

Think of Another Question

What else would you like to find out about what happens to an object's temperature when light shines on it? How could you find an answer to this new question?

In some restaurants, lights are used to keep food warm.

Investigate Light and Objects

What happens to light when it shines on different objects?

Science Process Vocabulary

compare verb

You **compare** when you say how things are alike and how they are different.

You can compare the results of your investigation with another group's.

When I compare the information, I see that both groups had the same results.

Materials

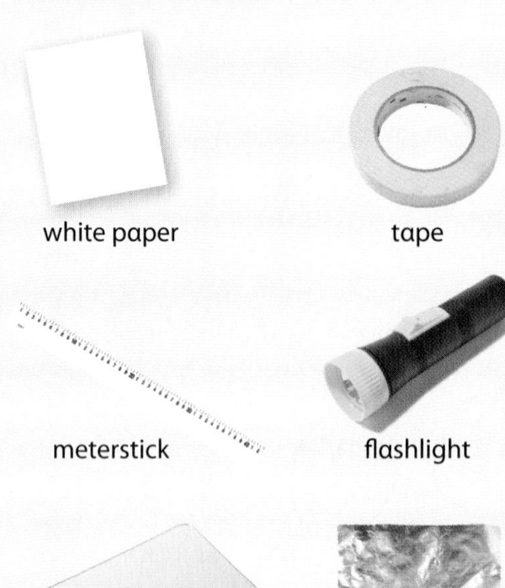

white paper

tape

meterstick

flashlight

mirror

foil

black paper

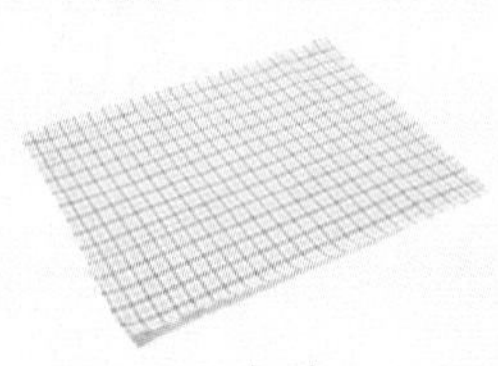

cloth

paper towel

What to Do

1. Tape a piece of white paper to the wall at eye level. **Measure** a distance of 50 cm from the paper. Put a piece of tape at that distance.

2. Stand at the piece of tape. Turn on the flashlight. Shine the flashlight on the white paper. **Observe** the light on the paper. Record your observations in your science notebook. Turn off the flashlight.

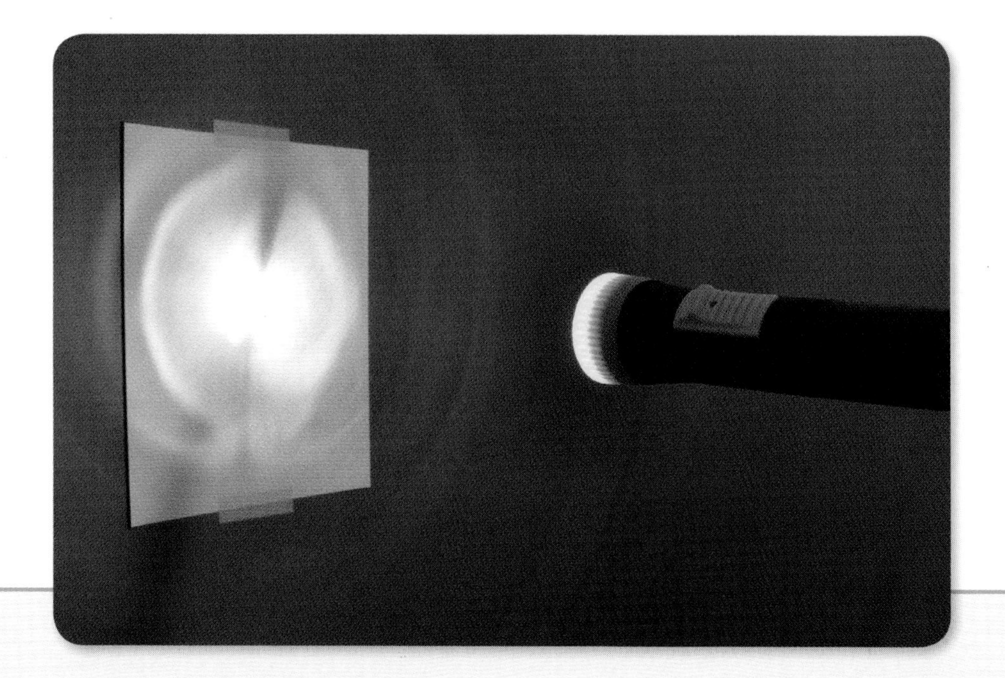

What to Do, continued

3. Select 3 objects. **Predict** what will happen to the light on the white paper when you put each object between the flashlight and the paper. Record your predictions.

4. Have a partner hold 1 object in front of you. Hold the flashlight behind the object, pointing it toward the white paper.

5. Turn on the flashlight. Observe the amount of light on the white paper and the object. Record your observations. Repeat with the other 2 objects.

Record

Write in your science notebook.
Use a table like this one.

What Happens to Light?

Material Between Flashlight and Paper	Predictions	Observations	
	On the white paper	On the white paper	On the object
None			

Explain and Conclude

1. **Compare** how bright the light on the white paper was when you shined it at the 3 different objects.
2. Were you able to see the same amount of light on all 3 objects in step 5? Why do you think that is so?
3. Based on the results of your investigation, what can you **infer** about what happens to light when it shines on different objects?

Think of Another Question

What else would you like to find out about what happens to light when it shines on different objects? How could you find an answer to this new question?

Open Inquiry

Do Your Own Investigation

Choose one of these questions, or make up one of your own to do your investigation.

- How do the masses of different pennies compare?
- How can you make water evaporate more quickly?
- How does changing the tightness of a piece of fishing line affect the pitch of the sound it makes?
- How can you line up mirrors to make light reflect on a certain spot?

Science Process Vocabulary

question noun

You ask a **question** to find out about something.

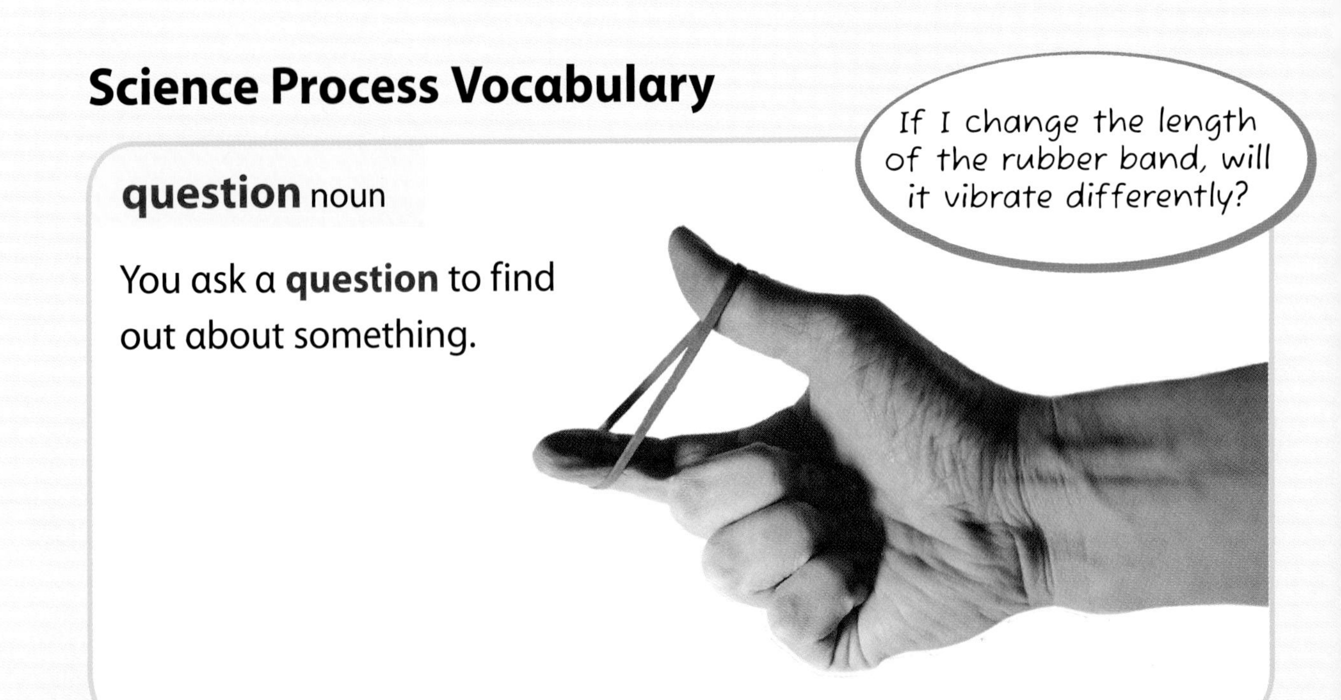

Open Inquiry Checklist

Here is a checklist you can use when you investigate.

Next Generation Sunshine State Standards
SC.3.N.1.1 Raise questions about the natural world, investigate them individually and in teams through free exploration and systematic investigations, and generate appropriate explanations based on those explorations. (Also **SC.3.N.1.3**)

- ☐ Choose a **question** or make up one of your own.
- ☐ Gather the materials you will use.
- ☐ If needed, make a **hypothesis** or a **prediction.**
- ☐ If needed, identify, manipulate, and control **variables.**
- ☐ Make a **plan** for your **investigation.**
- ☐ Carry out your **plan.**
- ☐ Collect and record **data. Analyze** your data.
- ☐ Explain and **share** your results.
- ☐ Tell what you **conclude.**
- ☐ Think of another question.

A siren has a high pitch.

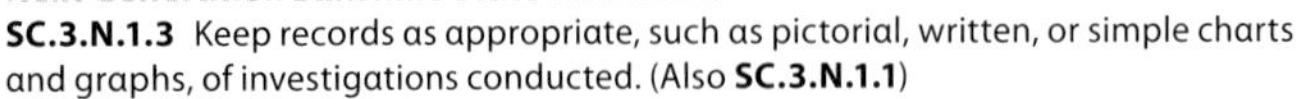
Next Generation Sunshine State Standards
SC.3.N.1.3 Keep records as appropriate, such as pictorial, written, or simple charts and graphs, of investigations conducted. (Also **SC.3.N.1.1**)

Write About an Investigation

Vibrations and Pitch

The following pages show how one student, Neal, wrote about an investigation. As Neal watched a musician tune his guitar, he wondered how changing the tightness of the guitar strings affected the sound the guitar made. He decided to investigate. Here is what he thought about to get started:

- Neal thought that changing the tightness of the guitar string would change the pitch of the sound the string made when plucked. He decided to do an investigation that showed how the tightness of a string affects pitch.
- He wanted to use simple materials. He would base his question and steps on the materials he would use.
- He decided to make a model guitar with a pegboard and fishing line.
- He would attach the pieces of fishing line to the pegboard with bolts and washers. He would change the tightness of the fishing line.

Model

Question

How does changing the tightness of a piece of fishing line affect the pitch of the sound it makes?

Choose a question that can be answered using materials that are safe and easy to obtain.

Materials

pegboard

fishing line

2 bolts

2 nuts

2 washers

List exact amounts of materials when needed.

Your Investigation

Now it's your turn to do your investigation and write about it.
Write about the following checklist items in your science notebook.

- [ ] **Choose a question or make up one of your own.**
- [ ] **Gather the materials you will use.**

Model

My Hypothesis

If I tighten the fishing line, then the sound it makes will have a higher pitch.

You can use "If…, then…." statements to make your prediction clear.

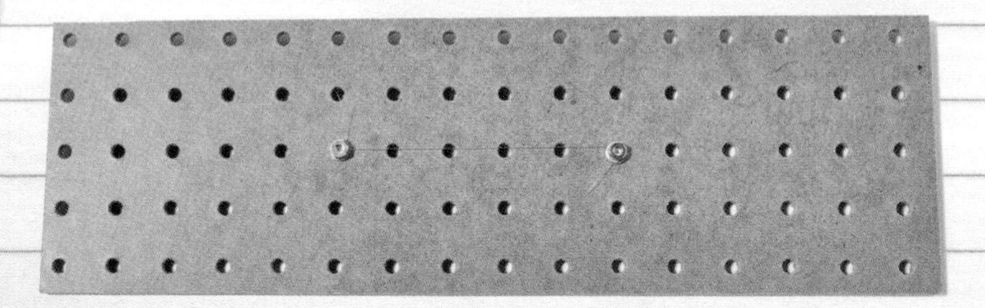

Your Investigation

☐ **If needed, make a hypothesis or prediction.**

Write your hypothesis or prediction in your science notebook.

Model

Variable I Will Change

I will change how tight the fishing line is stretched over the pegboard.

Variable I Will Observe

I will observe the pitch of the sound the fishing line makes when it is plucked.

Variables I Will Keep the Same

Everything else will be the same. I will pluck the fishing line with the same amount of force each time. The length of the fishing line will be the same each time.

Answer these three questions:
1. What one thing will I change?
2. What will I observe or measure?
3. What things will I keep the same?

Your Investigation

- [ ] **If needed, identify, manipulate, and control variables.**

Write about the variables for your investigation.

Model

My Plan

1. Place the washer on the bolt. Place the washer and bolt through a hole at one end of the pegboard. The top of the bolt should be on top of the pegboard.
2. Tighten the nut on the bolt.
3. Tie a piece of fishing line around the bolt. Pull the fishing line as tight as possible.
4. Tie the other end of the fishing line to another bolt, washer, and nut on the other end of the pegboard.
5. Turn the bolt to tighten the fishing line. Pluck it and listen to its pitch.
6. Tighten the fishing line again. Pluck it to see how the pitch changes.
7. Repeat step 6.

Write detailed plans. Another student should be able to repeat your investigation without asking any questions.

Your Investigation

☐ **Make a plan for your investigation.**

Write the steps for your plan.

Model

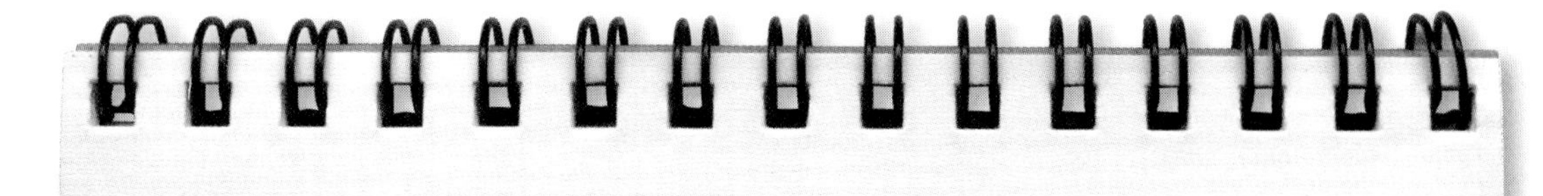

I carried out all 7 steps of my plan

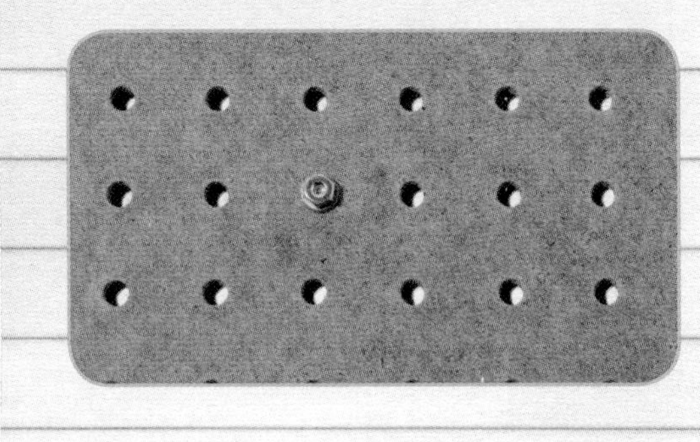

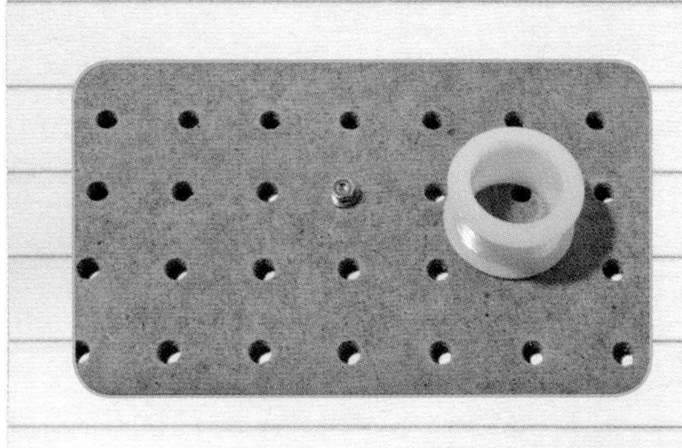

You might make a note if you needed to adjust your plan in any way. Neal found that he did not need to adjust his plan.

Your Investigation

☐ **Carry out your plan.**

Be sure to follow your plan carefully.

Model

Data (My Observations)

Pitch of String

	Tightness of Fishing Line	Pitch of Sound
Trial 1	loose	low
Trial 2	tighter	higher
Trial 3	tightest	highest

My Analysis

The fishing line made a sound with a higher pitch when it was stretched tighter.

Explain what happened based on the data you collected.

Your Investigation

☐ **Collect and record data. Analyze your data.**

Collect and record your data, and then write your analysis.

Model

Scientists often share results by demonstrating something.

Tell what you conclude and what evidence you have for your conclusion.

Investigations often lead to new questions for Inquiry.

How I Shared My Results

I played my model guitar for the class. I demonstrated what happened to the pitch of the fishing line when I changed its tightness. Then I shared my conclusions.

My Conclusion

As the fishing line is tightened, the pitch becomes higher.

Another Question

I wonder what would happen if I used strings of different thicknesses. Would the pitch change the same way?

Your Investigation

- [ ] Explain and share your results.
- [ ] Tell what you conclude.
- [ ] Think of another question.

Next Generation Sunshine State Standards
SC.3.N.1.7 Explain that empirical evidence is information, such as observations or measurements, that is used to help validate explanations of natural phenomena.

How Scientists Work

Using Observations to Evaluate Explanations

Scientists test their ideas by doing investigations. They make observations and gather data about the natural world. Then they analyze and interpret the data to explain their ideas.

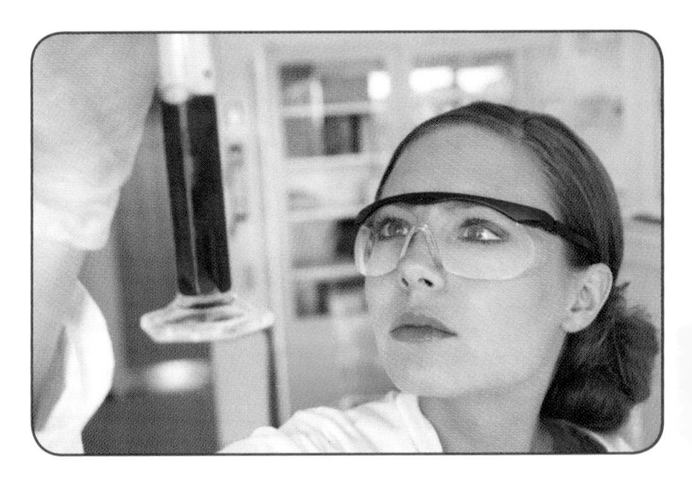

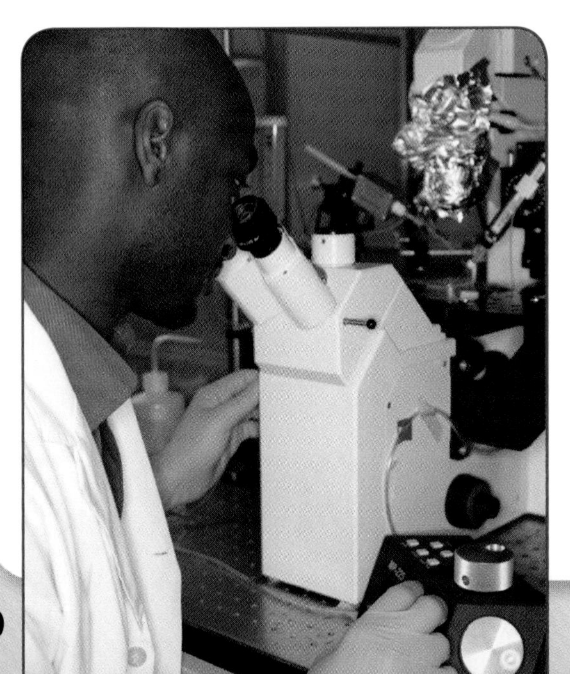

Making Observations and Collecting Data When scientists investigate, they gather information, called data. Data can be gathered by seeing, hearing, smelling, tasting, or feeling something.

Recording and Organizing Data Scientists might record data as written descriptions of what they observe. For example, a scientist might describe the color of an object or its motion. They might describe how an animal acts.

Other times data might be recorded as drawings or photographs. If scientists use measuring tools to make observations, their data might be numbers.

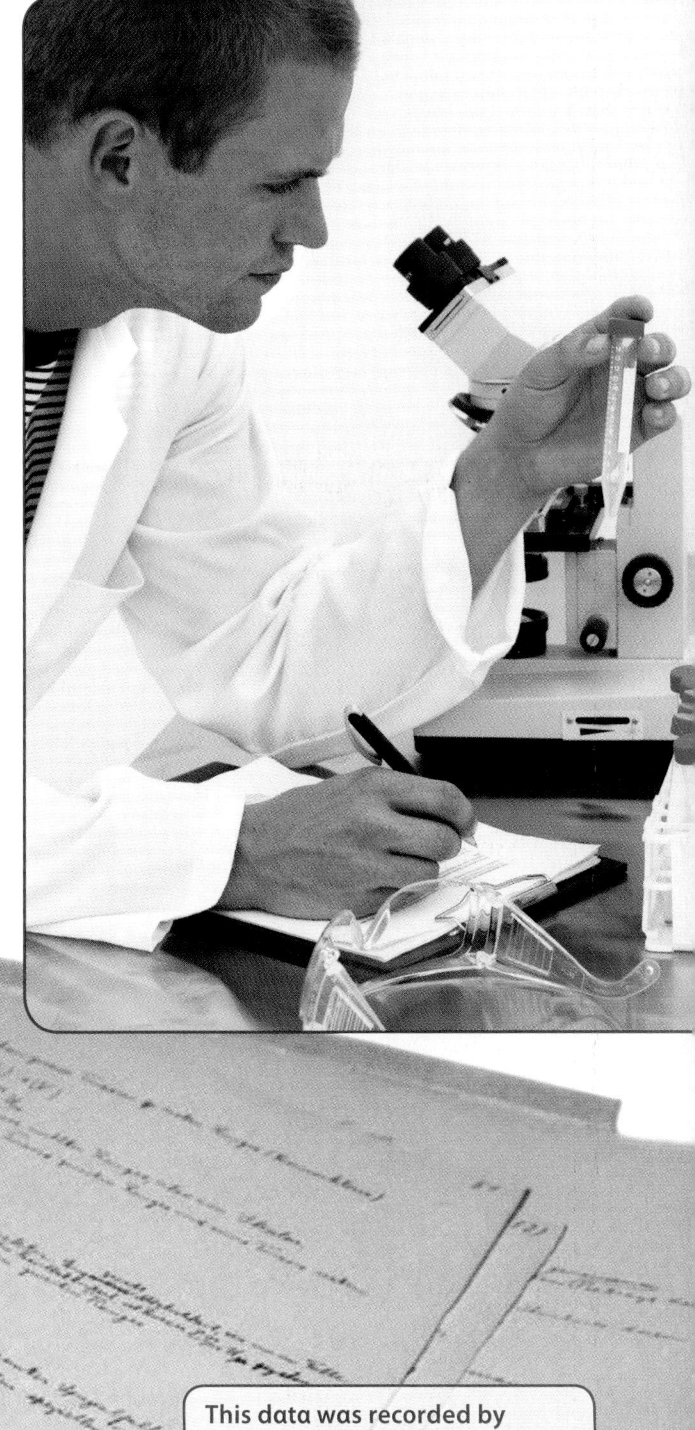

This data was recorded by famous scientist Albert Einstein.

Analyzing Data Then scientists analyze their data. They look for patterns. They decide whether the data support their ideas and explanations.

Sometimes the results of an investigation may not support a scientist's ideas. Then the scientist might ask questions such as:

- Did I make mistakes?
- Do I need to collect more data?
- What else do I need to know?
- How can I change my ideas so that they are supported by the new data?

SUMMARIZE

What Did You Find Out?

1. What are three ways that scientists might organize their data?
2. Why do scientists analyze their data?

Evaluate Explanations

Rico observed 2 water balloons of different sizes. He thought that the bigger water balloon would have more mass because it held more water. He decided to do an investigation to find out.

First Rico used a balance to measure the mass of the small water balloon. Then he measured the mass of the big water balloon. Rico measured each balloon several times to be sure that his measurements were accurate. He organized his data in a table.

Mass of Water Balloons

Trial	Mass of Large Balloon	Mass of Small Balloon
1	615 g	246 g
2	615 g	246 g
3	615 g	246 g

Analyze Rico's data. Write a paragraph telling whether you agree with Rico's idea that the bigger water balloon has more mass. Use Rico's data to support your argument.

ACKNOWLEDGMENTS
Grateful acknowledgment is given to the authors, artists, photographers, museums, publishers, and agents for permission to reprint copyrighted material. Every effort has been made to secure the appropriate permission. If any omissions have been made or if corrections are required, please contact the Publisher.

ILLUSTRATOR CREDITS
All maps by Mapping Specialists.

PHOTOGRAPHIC CREDITS
Front, Back Cover, 1–2, 3–4, 5 Image Quest Marine/Alamy Images. **6** DigitalStock/Corbis. **9** (tl) Tony Campbell/Shutterstock. (tc) Anton Foltin/ Shutterstock. (tr) Rui Saraiva/Shutterstock. **10** (c) Dmitry Naumov/Shutterstock. (b) Thomas M Perkins/Shutterstock. **10–11** (t) PhotoDisc/Getty Images. **12** (t) PhotoDisc/Getty Images. **13** irishman/Shutterstock. **14** (c) Hill Street Studios/ Blend Images/Alamy Images. (tbl) János Gehring/Shutterstock. (blc) bojan fatur/ iStockphoto. (tbr) Debra McGuire/iStockphoto. (bl) Socrates/Shutterstock. (br) Dominator/Shutterstock. **14–15, 16–17** (t) Andrew Williams/Shutterstock. **15** tl) János Gehring/Shutterstock. (tr) Kei Uesugi/Stone/Getty Images. (cl) Nigel Cattlin/Photo Researchers, Inc.. (cr) Nigel Cattlin/Alamy Images. (bl) Alexander Hafemann/iStockphoto. (bc) Lyroky/Alamy Images. (br) John Foxx Images/ Imagestate. **17** (tl) Debra McGuire/iStockphoto. (tr) Sally Scott/Shutterstock. (cl) Phil Schermeister/National Geographic Image Collection. (cr) Habman_18/ Shutterstock. (bl) Dominator/Shutterstock. (br) Peter Chadwick/Dorling Kindersley/Getty Images. **18** (t) Andrew Williams/Shutterstock. (bl) Socrates/ Shutterstock. (bc) bojan fatur/iStockphoto. (br) Praseodimio/Shutterstock. **20** (t) Cheryl Casey/Shutterstock. (bl) Amar and Isabelle Guillen - Guillen Photography/Alamy Images. (bc) stephan kerkhofs/Shutterstock. (br) Michael Patrick O'Neill/Photo Researchers, Inc.. **21** James Randklev/Corbis. **22** (t) Aaron S. Fink/age fotostock. (b) Karl H. Switak/Photo Researchers, Inc.. **23** (t) Cheryl Casey/Shutterstock. (l) Jim McKinley/Alamy Images. (cr) Creatas/Jupiterimages. (br) Johner Images/Alamy Images. **24** (b) Ace Stock Limited/Alamy Images. **24–25, 26** (t) Bianca Lavies/National Geographic Image Collection. **25** (clockwise l-r) Nick Biemans/Shutterstock. creacart/iStockphoto. Morozova Tatyana (Manamana)/Shutterstock. Dennis Sabo/iStockphoto. tom viggars/ Alamy Images. Joe Belanger/Shutterstock. Sebastian Duda/Shutterstock. Ivanova Inga/Shutterstock. Zuzule/Shutterstock. Digital Vision/Getty Images. **26** (c) Digital Vision/Getty Images. **27** (b) Frederick R. Matzen/Shutterstock. 28 (tl) Awei/Shutterstock. (c) EcoPrint/Shutterstock. (tbl) Kyle Ward/National Geographic Image Collection. (tbc) Ramesh Patil/National Geographic Image Collection. (tbr) Darylne A. Murawski/National Geographic Image Collection. (bl) Tim Laman/National Geographic Image Collection. (bc) Darlyne A. Murawski/ National Geographic Image Collection. (br) Tom Vezo/Minden Pictures/ National Geographic Image Collection. **29** (tr) Nekan/iStockphoto. **30** (tl) Awei/ Shutterstock. (tr) Nekan/iStockphoto. **31** (bl) Arco Images GmbH/Alamy Images. (br) Ingram Publishing/Superstock. **32, 34** (t) kwest/Shutterstock. (b) JGI/Blend Images/Getty Images. **35** (b) ex0rzist/Shutterstock. **36, 38** (t) Anette Linnea Rasmussen/Shutterstock. **39** (b) DigitalStock/Corbis. **40** (b) Nancy Kennedy/ Shutterstock. **40–41** (t) Els Jooren/Shutterstock. **41** (b) Chris Curtis/Shutterstock. **42** (t) Wong Hock weng/Shutterstock. (b) chudoba/Shutterstock. **44, 46, 49** (t) Wong Hock weng/Shutterstock. **46** (t) Wong Hock weng/Shutterstock. **50** (t) Jonathan Blair/Corbis. (b) dubassy/Shutterstock. **51** (bg) David Gralian/Alamy Images. (t) John Cancalosi/Alamy Images. (b) Arco Images GmbH/Alamy Images. **52** (t) James L. Amos/National Geographic Image Collection. (b) Joe Tucciarone/ Photo Researchers, Inc.. **53** (t) Jonathan Blair/Corbis. (c) Arpad Benedek/ iStockphoto. (b) Ralph Lee Hopkins/National Geographic Image Collection. **54** (l) Orange Line Media/Shutterstock. **55** (l) PhotoDisc/Getty Images. (r) Shioguchi/ Getty Images. **56** (bl) PhotoDisc/Getty Images. (br) MetaTools. **56–57, 58** (t) Brand X Pictures/Jupiterimages. **59** (b) Ingrid Balabanova/Shutterstock. **60** (t) Chernetskiy/Shutterstock. **61** Mark Scheuern/Alamy Images. **63** (t) Chernetskiy/ Shutterstock. (b) Ingram Publishing RF/Photolibrary. **64, 66** (t) Kezzu/ Shutterstock. (b) Emory Kristof/National Geographic Image Collection. **67** (b) DigitalStock/Corbis. **68** (b) Mark Andersen/Getty Images. **68–69** (t) DigitalStock/ Corbis. **71** (b) PhotoDisc/Getty Images. **72** (t) NASA/JPL-Caltech/S. Carey (SSC/ Caltech)/JPL (NASA). (c) amana images inc./Alamy Images. (bl) Jim Sugar/ Corbis. (bc) Germany Feng/Shutterstock. (br) Intraclique LLC/Shutterstock. **74** (t) NASA/JPL-Caltech/S. Carey (SSC/Caltech)/JPL (NASA). **76–77** (t) image 100. **77** (b) moodboard/Alamy Images. **78–79, 80–81, 82–83, 84–85** (t) PhotoDisc/ Getty Images. **86** (t) Serg64/Shutterstock. (c) SOHO (ESA & NASA). (b) Brand X/ Corbis. **87** Brooks Kraft/Corbis. **88** Tina Stallard/Getty Images. **89** (t) Serg64/ Shutterstock. **94–95, 96** (t) Malcolm Leman/Shutterstock. **97** (b) Igor Zhorov/ iStockphoto. **98** (c) Jani Bryson/iStockphoto. **98–99** (t) Jupiterimages. **101** (b) James Steidl/Shutterstock. **102** (t) Gerry Ellis/Minden Pictures. (b) James P. Blair/ National Geographic Image Collection. **104** (t) Gerry Ellis/Minden Pictures. **105** (b) Stockbyte/Getty Images. **106** (t) Ian Shaw/Alamy Images. (c) Julien/ Shutterstock. (b) Science Photo Library/Alamy Images. **107** (l) Jack schiffer/ Shutterstock. **109** (bg) John Foxx Images/Imagestate. (t) Ian Shaw/Alamy Images. **110** (t) Nikita Rogul/Shutterstock. (b) John Foxx Images/Imagestate. **112** (t) Nikita Rogul/Shutterstock. **113** (b) Scientifica/Getty Images. **114** (t) Brent Walker/Shutterstock. (c) Ilin Sergey/Shutterstock. **116** (t) Brent Walker/ Shutterstock. **117** (b) Bruce Bean/iStockphoto. **118, 120** (t) ccc/Shutterstock. **121** (b) Peter Albrektsen/Shutterstock. **122** (b) malcolm romain/iStockphoto. **122–123, 124** (t) saied shahin kiya/Shutterstock. **125** (b) Yu Zhang/iStockphoto. **126** (t) nikkytok/Shutterstock. (b) bonnie jacobs/iStockphoto. **128** (t) nikkytok/ Shutterstock. **129** (b) Tokar Dima/Shutterstock. **130** (t) Aron Hsiao/Shutterstock. (b) hd connelly/Shutterstock. **131** (t) Aron Hsiao/Shutterstock. (b) David Touchtone/Shutterstock. **132** (b) Premier Edition Image Library/Superstock. **132–133, 134–135, 136–137, 138–139** (t) Steve Cole/PhotoDisc/Getty Images. **140** (t) PhotoDisc/Getty Images. (c) Yuri Arcurs/Shutterstock. (b) Digital Vision/ Noel Hendrickson/Getty Images. **141** (t) Wouter Tolenaars/Shutterstock. (b) Jon Levy/Getty Images. **142** Laurence Gough/iStockphoto. **143** (t) PhotoDisc/ Getty Images.

PROGRAM AUTHORS
Judith Sweeney Lederman, Ph.D., Director of Teacher Education and Associate Professor of Science Education, Department of Mathematics and Science Education, Illinois Institute of Technology, Chicago, Illinois; Randy Bell, Ph.D., Associate Professor of Science Education, University of Virginia, Charlottesville, Virginia; Malcolm B. Butler, Ph.D., Associate Professor of Science Education, University of South Florida, St. Petersburg, Florida; Kathy Cabe Trundle, Ph.D., Associate Professor of Early Childhood Science Education, The Ohio State University, Columbus, Ohio; David W. Moore, Ph.D., Professor of Education, College of Teacher Education and Leadership, Arizona State University, Tempe, Arizona

National Geographic School Publishing
Hampton-Brown
www.NGSP.com

Printed in the USA.
RR Donnelley
Menasha, WI

ISBN: 978-0-7362-7734-1

11 12 13 14 15 16 17 18 19 20

3 4 5 6 7 8 9 10